PARTNERSHIP AND LEADERSHIP

ECO-EFFICIENCY IN INDUSTRY AND SCIENCE

VOLUME 8

The titles published in this series are listed at the end of this volume.

Partnership and Leadership

Building Alliances for a Sustainable Future

Edited by

Theo J.N.M. de Bruijn
Center for Clean Technology and Environmental Policy,
University of Twente,
Enschede, The Netherlands

and

Arnold Tukker
TNO Institute of Strategy, Technology and Policy,
Delft, The Netherlands

KLUWER ACADEMIC PUBLISHERS
DORDRECHT / BOSTON / LONDON

A C.I.P. Catalogue record for this book is available from the Library of Congress.

ISBN 978-90-481-5938-3

Published by Kluwer Academic Publishers,
P.O. Box 17, 3300 AA Dordrecht, The Netherlands.

Sold and distributed in North, Central and South America
by Kluwer Academic Publishers,
101 Philip Drive, Norwell, MA 02061, U.S.A.

In all other countries, sold and distributed
by Kluwer Academic Publishers,
P.O. Box 322, 3300 AH Dordrecht, The Netherlands.

Printed on acid-free paper

TABLE OF CONTENTS

PART 2: COLLABORATION ALONG THE VALUE CHAIN

PART 3: MULTISTAKEHOLDER COLLABORATION

Chapter 1

INTRODUCTION

Collaboration as the New Paradigm in Industrial Transformation

THEO DE BRUIJN

ARNOLD TUKKER

1. BACKGROUND OF THE BOOK PROJECT

Sustainable development has become a central perspective in environmental strategies around the world. In order to reach for sustainable development fundamental changes are needed in current production and consumption structures. It is clear that governments nor businesses have the capacity to engage society in such a transformational process on their own. Therefore, collaboration has emerged as a central concept in many strategies. At the same time it is obvious that someone has to take the lead in the development towards sustainability. Thus, partnership and leadership are strongly related concepts. This book describes different forms of collaboration emerging between various actors in order to transform the industrial production system. The objective of the book is to systematically explore the different roles and relationships between partnership and leadership. The main question we put forward is to what extent they contribute to the process of sustainable development.

The idea of this book was born at the seventh International Conference of the Greening of Industry Network that took place in Rome, Italy, November 15-18, 1998. This conference focused on transformation processes and the

T.J.N.M. de Bruijn and A. Tukker (eds.), Partnership and Leadership, 5–20.

roles of different actors in such processes. The main theme of the conference was the engagement of partners in the transition of society towards sustainability. This transformation requires the co-operation, perseverance and dedication of all parties involved, and the development of new procedures, institutions and cultures needed for a productive co-operation. Options for this and routes towards it were explored during plenary sessions, breakout sessions and workshops devoted to the building of alliances in a global, regional and local setting.

Next to a conceptual part, the book brings together case studies from around the world. Its focus is on describing and understanding various formats of collaboration and critically evaluating its effects and prospects. The book is in part an edited version of selected papers presented during the GIN conference in Rome on the subject, enriched with contributions of other invited authors.

2. COLLABORATION AND INDUSTRIAL TRANSFORMATION

Since the late 1960s major changes have taken place in environmental strategies. Where initially environmental action consisted of isolated activities of minor concern, it has now developed into a field for strategic action by both governments and business. The issue at stake has changed drastically: from taking care of local nuisance problems to global problems that are strongly related to economic and social issues. Ever since the publication of the report *Our common future* (World Commission on Environment and Development - Commission Brundtland, 1987) and the United Nations Conference on Environment and Development (UNCED) in Rio de Janeiro in 1992, sustainable development has become the central perspective in national and international environmental strategies. Modern approaches aim at providing basic human needs by ensuring economic growth within ecological boundaries. Important changes in production and consumption systems are required in order to meet the needs and aspirations of a growing world population while using environmental resources in a sustainable manner (IHDP Industrial Transformation Science Plan 1999: xi). The challenge is to develop strategies through which society can combine economic and social development with the reduction of pressure on the environment. De-linking economical development from its negative external effects such as environmental pressures is essential. Obviously, such strategies exceed the capacity of individual actors. Both governments and industry therefore embrace the concept of collaboration.

To illustrate the governmental perspective we can quote former European Commissioner Ritt Bjerregaard who stated that *"environmental regulation of industry needs to move away from confrontation towards a new phase of co-operation and trust"* (Bjerregaard, 1997). Here the former commissioner refers to one of the key principles of the Fifth Environmental Action Program of the European Union (CEC, 1993), namely shared responsibility. Shared responsibility was born out of a notion that governments need the aid from industry, NGOs and the general public to bring about the changes needed. Throughout the world we witness government policies moving away from confrontation toward more collaboration (De Bruijn and Norberg-Bohm, forthcoming). When environmental degradation emerged as a priority for government action in the early 1970s, most countries enacted media-specific legislation based on direct regulation (i.e. ambient, emission and technology standards enforced through permitting systems). Although direct regulation can be a powerful tool for adjusting industrial behavior, it is less adequate to address the challenges of sustainable development. Sustainable development is about more than 'compliance to regulation'. Especially when there is so much uncertainty as with sustainable development, it goes beyond the capacity and capability of governments to develop meaningful and effective legislation. Governments increasingly realise that industry is not only part of the problem, but also part of the solution through the development of new processes, technologies and products (CEC, 1993). Key elements in modern environmental policy are therefore collaboration and interaction (Hartman et. al., 1999). New approaches try to strengthen the dialogue and consultation with concerned parties (Lévêque, 1996). Creating interaction among the main groups of actors including governments, public and private industry and the general public is seen as one of the paths towards sustainable development.

Not only governments opt for more collaboration. Industry itself has been moving in same direction during the last decade. A famous statement is by Loudon (1987), former president-director of AKZO Nobel: *"An industry cannot survive an ongoing conflict with society. Dialogue, adjustment and collaboration are not a luxury but a necessity"*. When the environment emerged as a theme for action in the early 1970s the initial response by industry was one of reluctance and denial (Fischer and Schot, 1993). This position soon proved to be untenable. Environmental policies were not a temporary phenomenon, neither were environmental concerns in society. Since the mid 1980s industry started to realise that environmental problems are part of their own responsibility (Schot, 1997). Industry also realised that a pro-active approach could pay off in the end: taking an active role in developing new measures is probably more beneficiary than just waiting for additional legislation that becomes available at unexpected and awkward times. This legislation was also developed by people (governments, regulators) without inside knowledge of industrial practices. A more active involvement can result in legislation that fits better with industrial realities. Furthermore, in the 1990s industry realised it could no longer legitimise its actions solely on the basis of governmental permission. That was the hard lesson that for instance Shell learned with the Brent Spar (Grolin, 1997). Although the British government had approved the dumping of the oil platform in the ocean, in the end Shell was forced to develop another plan by societal protests lead by Greenpeace. Communication and collaboration with a broader circle of stakeholders has therefore become a necessity. Quoting Björn Stigson (President of the World Business Council for Sustainable Development): *"By cooperating with a wide range of stakeholders, we can concur toward common objectives for sustainable development, and make tangible progress toward achieving them"* (Stigson, 2001).

Both governments and industry thus embrace collaboration as a promising strategy in pursuing towards sustainable development. Nevertheless, the concept of sustainable development presents an enormous challenge. For one, companies will have to become much more 'eco-efficient'. Eco-efficiency means 'doing more with less' (see for instance Schmidheiny, 1992; Von Weizsäcker et al., 1997). Eco-efficiency therefore implies de-materialisation, maximising the use of renewable inputs, minimisation of the use of energy and increasing the ability to recycle products (Schot, 1997). And indeed, after a decade of intensifying the goals and implementation of environmental policy a lot of companies have managed to improve their environmental efficiency, especially in developed countries (De Bruijn et al., 1997). To become more eco-efficient companies need too develop stakeholder relationships (Schot et al., 1997). Perceived benefits of these relationships are that stakeholders can deliver 'fresh' and new ideas, they can provide support for corporate activities and they can ensure acceptance and successfulness of corporate strategies in other parts of the life-cycle of the companies' products.

Industrial transformation however goes beyond the notion of eco-efficiency and beyond the domain of individual actors. It is about system innovation, both technological and institutional[1]. Economic growth nullifies much of the benefits achieved with eco-efficiency and more efficiency will therefore not be sufficient (see for instance Clark and George, 1995; Huber, 1995). Eco-efficiency makes the current, destructive system less so. It slows

[1]In chapter 16 we will further classify innovations in system optimisation, singular innovation, and system innovation.

down the rates of contamination and depletion, but does not stop these processes (McDonough and Braungart, 1998). Moreover, eco-efficiency does not pay attention to social dimensions of sustainable development (for instance inter- and intra-generational equity). The real challenge for the coming years is in forcing structural changes in the production and consumption system, at a higher level than individual actors. Roome (2001:3) states in this respect that "given the serious overuse of materials and resources arising as a result of our developed industrial life-styles and population pressures, the innovations demanded by any transition to sustainable development are likely to be at the more complex end of the hierarchy". In this hierarchy he distinguishes between 'compliance', 'beyond-compliance', and the 'sustainable enterprise'. He also states that "moving up the innovation hierarchy increases in the complexity of the issues, the numbers of actors involved in change, and the number of linked, multiple technological and social options, the innovations and new practices that need to be undertaken and the uncertainties that have to be considered". Where companies can interact mainly with regulators, customers and neighbours during the phase of compliance, sustainable enterprises need the involvement of all actors in production and consumption systems. McDonough and Braungart (1998) speak in this respect about 'the next industrial revolution', a new way of designing industrial production. Instead of becoming more efficient they argue for new designing principles, such as 'waste equals food' and 'cradle-to-cradle'[2].

Obviously, industrial transformation cannot be planned by a single actor or a single sector. It even involves geographic scales that transcend a single country or a single continent (IHDP Industrial Transformation Science Plan, 1999: xii). Mutual adjustment and learning by a variety of actors are important strategies in this respect. Industrial transformation therefore requires collaboration at various levels, e.g. among industrial sectors and societal sectors in a North – South perspective (Hartman et al., 1999).

In sum, there are several benefits and motivations for increased collaboration. First, collaboration can bring new perspectives and knowledge to (policy and corporate) processes, resulting in better solutions than parties can create individually. Second, the challenges in sustainable development and industrial transformation exceed the capacity of individual actors. Via multi-actor processes visions of the future and more sustainable socio-

[2]They distinguish between biological and technical metabolism. Within the biological metabolism products and materials have to break down under natural conditions. They are designed to return to the organic cycle ('waste equals food'). Products composed of materials that do not biodegrade should be designed as technical nutrients that circulate within closed-loop industrial cycles (cradle-to-cradle).

technical systems can be developed and discussed (Roome and Cahill, 2001). Governments are looking to collaboration as a way of mobilising the knowledge and creativity of industry. For industry, collaboration on environmental matters has become a reality over the last decade due to an increased scrutiny and interaction with a broad range of stakeholders including environmental interest groups, community groups, shareholders, boards of directors, insurance companies and investment companies.

Despite the wealth of arguments in favor of collaboration and the increasing practice of parties coming together, there are also worries. The main worry relates to the issue of radical change. For industry to make the kind of changes as in industrial transformation will require severe pressures. Critics fear that collaboration will take off some of the 'heat'. Hartman et al. (1999) state that partnerships and collaboration appear to favor incremental progress rather than more fundamental changes. It might be the case that collaboration doesn't lead to sharing responsibility but to situations in which no actors feel responsible any more. *Sharing* responsibility is then confused with *dividing* responsibility. Also, in collaborative settings someone needs to give direction to the transformation process. Basically the question is where the imperative for change is in collaborative settings, who will provide the leadership that is needed. It cannot be expected that there is a natural tendency of societal actors to 'align' their actions in such a way that they will co-operate smoothly to realise a transition. Hartman et al. (1999) argue for leadership without recurring to hierarchical forms of leadership. They identify what they call collaborative leadership (see also chapter 2 in this volume). Collaborative leaders initiate collaboration by bringing the right actors together and ensure that the process is sustained. Without forcing their perspective to the collaborative process they keep it going and progressing toward common shared goals. Finally and related to the second issue is the question of *how* parties can collaborate when changes require the involvement of multiple actors in an international setting. There is still a lot of uncertainty on the kind of methods that can be applied successfully.

3. FOCUS OF THE BOOK

Collaboration has emerged as the central question today in processes of industrial transformation. Although it is widely recognised as a necessity the design of *effective* collaborative settings is still much debated. This book aims at taking stock of the experiences so far with partnerships and leadership. It aims at explaining, mapping and analysing different scopes/goals/formats/methods of collaborative approaches, and at the same time it takes a critical approach to the theme by understanding related risks, side-effects, and prospects. The different chapters in this book introduce important concepts and issues concerning collaboration, and present numerous cases from around the world of processes in practice in which parties are coming together.

Let's first define what we mean by collaboration, partnership and leadership. In defining collaboration we follow Gray (1989): "a process through which parties who see different aspects of a problem can constructively explore their differences and search for solutions that go beyond their own limited vision of what is possible". Collaboration therefore involves actors from different positions. We focus on two important aspects in collaboration: partnership and leadership. Partnerships can be defined as "voluntary collaborations between two or more organisations with a jointly-defined agenda focused on a discrete, attainable and potentially measurable goal" (Long and Arnold, 1995). Partnerships are therefore a specific form of collaboration, namely those in which actors share a common goal. As we have stated, actors cannot be expected to align their actions naturally. Some form of 'guidance' by a collaborative leadership is necessary. Collaborative leadership is defined by Hartman et al. (1999) as "the process to initiate, facilitate and sustain collaborative initiatives among stakeholders for addressing environmental sustainability issues". The chapters in this book focus on what the scope of the partnerships was, what goals it did pursue, with what effects and what roles the different partners played. Special attention is paid to the question of who took the lead in the partnership.

As we have claimed in the previous section, both partnership and leadership can be viewed in a positive and a negative way. On the one hand, sustainable development simply calls for more collaboration since the changes needed exceed the capacity of individual actors. But partnership and consensus-based decisions can also be seen as an obstacle to realizing radical changes in production and consumption structures. Similarly, leadership can be seen as an obstacle to sustainable development if leaders form close circles and are not willing to share experiences with other actors; but leadership is also seen as an important element to keep concepts and practices moving forward. Since the arguments of both the proponents and opponents seem to have some validity at least, the main question is not whether to strive for more collaboration or not, but rather how to design effective partnerships. In the first, conceptual part of the book some important issues are identified in this respect. The cases that follow will be used to assess the following questions: How does collaboration look like in practice? Who's leading the partnership? What are the effects?

We will explore these questions by looking at two different formats of collaboration. With Roome (2001) we acknowledge that innovations with different levels of ambition require different sets of actors to collaborate. As a first step, we may observe that sustainable development calls for fundamental changes that exceed the level of individual companies. The changes needed are only possible when other actors in the economic network (e.g. suppliers, customers) are engaged in the process. Several chapters in the book focus on collaborative attempts that involve different partners along the value chain, that is actors mainly from within the economic network of companies. A true industrial transformation to sustainable development, however, influences many more actors. Therefore, as a second step we focus on multi-stakeholder collaboration. These partnerships involve a broad category of actors including industry, governments, NGOs, research institutes, consumers, etc., often in an international perspective. For both kinds of partnerships we will assess its promises for industrial transformation.

4. INTRODUCTION TO THE CHAPTERS

The book consists of three parts. Part I is the conceptual part of the book. Part II focuses on increased collaboration along the value chain and part III explores multi-stakeholder partnerships.

In Part I different dimensions of collaboration are reflected upon, and the major issues around collaboration are identified. Together the chapters provide a framework of reference for the chapters that follow. The first two chapters in Part I focus on the potential value of collaboration. The first chapter in Part I is by *Cathy Hartman, Peter Hofman and Edwin Stafford*, a revision of a paper that was originally written as a reflection of the papers and debate of the GIN conference in Rome (Hartman et al., 1999). Hartman, Hofman and Stafford assess the value of collaboration in reaching for sustainability and identify positive expectations next to firm criticism. They identify several characteristics of collaboration that make it a highly suitable concept for addressing environmental problems and steering society toward sustainability. First, environmental problems are complex and actions must be taken without the benefit of conclusive evidence. Collaboration offers the opportunity to bring together experts from a variety of different disciplines and arenas to fashion solutions that can go beyond the limited perspectives of individual stakeholders. Second, environmental solutions often require a cadre of economic, political, social, and marketing resources. Collaboration creates a stakeholder network for the marshalling of necessary resources.

Third, environmental solutions typically bring new stakeholders to the table who must work together in novel ways. As opposed to traditional command and control or adversarial approaches, collaboration can work outside the formal political, social, and economic structures in society with the flexibility to build stakeholder networks relevant to the problem at hand. In short, collaboration is essential to sustainable development. Critics on the other hand state that collaboration tends to provide only incremental environmental progress rather than overt, discontinuous innovations or "systems change" because collaboration does not allow for sufficient changes in society's values and political structures necessary for sustainable development. They also note that in practice, collaboration has not been a panacea for environmental transformation. They argue that collaboration and conflict are two ends of a continuum that need to be employed simultaneously: Because sustainable development is not well defined, collaboration is needed to facilitate the constructive exploration of environmental stakeholder differences to establish collective understandings of the necessary path toward sustainability. The threat of conflict is needed to encourage dialogue and participation. So in the end, they conclude that rather than characterising traditional adversarial approaches and collaboration as either/or paths for sustainability, the two may actually go hand in hand in a synergistic fashion.

In his chapter, *Eric Poncelet* explores the particular implications that multi-stakeholder partnerships (involving business, governments and NGOs) have for the pursuit of sustainable development. His perspective focuses on the diverse manners by which different actors conventionally think, talk, and act with regard to specific environmental problems, and he argues that partnerships should be viewed as encounters between differentially valued meaning systems rather than merely as confrontations among competing interests. He applies this perspective to the study of two partnerships -a European Union (EU) level initiative aimed at enabling the practice of sustainable development in Europe, and a national-level project directed toward bettering natural resource management in the United States (U.S.)- where he finds one discourse to be favored above all others. This was the discourse of ecological modernisation, with its central concept of economic-ecological 'win-win' solutions. Poncelet concludes that partnerships promoting the ideology of ecological modernisation can be of value in reaching for sustainable development, although he notes that the focus on finding such 'win-win' solutions serves to keep some issues and ideas off the table.

Susse Georg and Alan Irwin introduce another issue concerning collaboration, namely the often heard contrast between local and global action. They note that the general rhetoric of sustainable development as expressed in the Brundtland Report emphasises the need for widespread participation and empowerment. Meanwhile, and somewhat counter to this approach, the process of institutional response and development has largely been dominated by national and inter-national bodies. Rather than emphasising the contrast, they point at the significance of local-global relations within particular environmental initiatives. Their central claim is networks generally combine the local and global rather than fitting neatly into either category. They review two cases – the development of an ecological village in Sweden and the development of CFC-free refrigerator. They show that both cases depend to some degree upon local networks and contacts but simultaneously they rely upon global relations of knowledge and expertise. They conclude that rather than advocating either the local or the global, we should be addressing their re-combination and re-construction within particular social and ecological settings.

The final two chapters in the first part each emphasise the need for learning processes in collaborative settings. *Nigel Roome and Sarah Clarke* state that environmental management and sustainable development require companies to acquire knowledge that is not ordinarily found in their existing repertoire or experience. Companies therefore need to participate in forms of collaborative action that link traditional business issues to a set of environmental and social concerns. They introduce the concept of a learning-action network, which they define as a set of "relationships which lay over and complement formal organisational structures linking individuals together by the flow of knowledge, information, and ideas". In such networks business organisations and stakeholders in society are engaged in continuous learning, action and change. To illustrate the meaning of the concept Clarke and Roome present the case-study of a Canadian company. In it they describe how the company's approach to strategic planning identified and responded to demands for more sustainable practices, how its approach was progressively refined and redefined, and the way that the organisation's culture and strategy processes influenced its willingness to learn and act with a network of internal and external stakeholders. Based on these findings they identify the critical role for learning-action networks in the transition to more sustainable business organisations.

Maj Munch Andersen focuses at inter-firm learning processes from an economic organisation theoretical perspective and asks how the greening of the economy is affected by the difficulties in organising the necessary coordination and learning processes within and notably between (asymmetrically greening) firms on the market. Do firms make each other go green as a part of the economic process? Are firms held back because some conservative firms or firms lacking green capabilities function as bottlenecks? Or are sectors, production chains, regions or other collections of firms swept forward together in a common collaborative green learning process, entailing a consolidation and assimilation of their (greener) strategies? A key theoretical question related to this discussion is whether the inter-firm greening process primarily is a collaborative or adversarial process. The chapter provides a conceptual framework for analysing inter-firm learning suggesting three different types of learning processes, illustrated by empirical examples of green inter-firm learning between two Danish paper mills and their suppliers, customers and competitors. The chapter concludes that system failures on the market are important factors shaping the greening process and that it is necessary to take into account the differences and interrelationships of different kinds of inter-firm learning processes. Inter-firm learning processes hence gain significance in efforts to understand retardation in industrial greening processes and ways of overcoming it.

Part II of the book focuses on collaboration between partners within the value chain. The development and employment of cleaner technologies, processes and products need collaboration that might go beyond the usual pattern of interaction. Part II of the book introduces case-studies that describe the development of those new interaction patterns. The collaboration involves suppliers, clients, trade associations, etc. In short, it concerns partners within the economic circle of the firm. The rational behind the interaction is to enable changes that would not take place on an individual basis. *Simone Maier* describes in her chapter the introduction and development of organic food in the Swiss market. She tells the story of one of the major Swiss retailers that tried to introduce an organic food line into its line of products. This process involved a labeling organisation and of course suppliers who could produce according to the requirements of the retailer. Although the favorable effect on the development of a market for organic food cannot be denied, Maier shows how difficult and frustrating the process can be for a single supplier, especially in situations where there is a large dissimilarity in power. She also claims that without some willingness on both sides the learning process necessary for mutual adjustments won't take place. From her case it is clear that the use of power alone (by the leader in the partnership) won't result in fruitful collaborative interaction patterns.

Louise Canning and Stuart Hanmer-Lloyd also state that the decision of either party to support the process is to some degree determined by an exchange partner's own interest in the eventual adaptation output as well as the perceived commitment, credibility and reliability of the other party. They base their conclusions on an exploration of the processes followed by companies to introduce environmental adaptations where the support of exchange partners is required. Based on four case-studies they conclude that environmental adaptations will occur as companies in an exchange relationship try to better align their activities in order to satisfy the environmental performance criteria of the respective parties and ensure the efficient use of resources. They are able to identify a number of factors that can hinder or facilitate the progression of the adaptation process. They especially call the attention to 'relationship management' in order to secure the support of partners. Canning and Hanmer-Lloyd find that the progression in some of the environmental adaptation processes exemplify that environmental improvements can not necessarily be seamlessly introduced into business relationships. Whilst some of these difficulties might derive from technical problems, the progression of the process is shown at times to be conditioned by the relationship between the two companies as well as the behavior and experience of those involved in the process. So trying to reduce the environmental impact of company activities requires the consideration of factors beyond the optimum technical solution being sought to include the effective management of the co-operative process between firms.

This conclusion is also being put forward by *Joanneke Kruijsen* in her analysis of the diffusion of Photo-Voltaic Technologies in the Netherlands. She states that the challenge of increasing the use of photovoltaic technologies is not a technological challenge only, it is also a social challenge. The new technology needs to be accepted by its social environment before it can develop into a substantial part of the sustainable or renewable energy supply mix. Her central conclusion therefore is that collaboration in the Dutch PV sector is a precondition for effective diffusion. None of the actors involved is able to stimulate the diffusion of photovoltaic technologies on their own. In her case she finds that the user was not always sufficiently involved. Market development is the missing factor in photovoltaic diffusion. Kruijsen gives some suggestions in her chapter how to get the market and users more involved.

The next two chapters both focus on business support networks in which governmental agencies, regional authorities, trade associations, consultants, etc. collaborate in an attempt to reach for and transform individual companies. *Frances Hines* describes an attempt in the UK to coordinate the support by an increasing number and diversity of organizations. This had led to a bewildering situation for company managers, who are often confused about the type of support available. The chapter describes a more coherent approach to environmental business support by the establishment of the Environmental Services Partnership. This partnership served as a single point of contact for SMEs and it provided coordination of funding applications for the participating organisations. While the model produced some benefits, mostly short-term direct benefits to business, it did not achieve its long-term aims of co-operation and co-ordination of environmental business support in Wales. The model did not change the fundamental structure or nature of relationships between participating organisations. Hines shows that previously competing organisations had severe problems in engaging themselves in a collaborating context. Although collaboration would lead to a better situation overall, it was hard for individual organisations to refrain from pursuing their own interests.

Theo de Bruijn and Peter Hofman also focus on support networks for SMEs. They state that SMEs not only need to be convinced of the necessity of changes but will also need concrete support in order to get engaged in learning processes necessary for sustainable development. Partnerships with external organisations might be able to deliver the environmental capabilities needed. De Bruijn and Hofman explore the effectiveness of these partnerships by focusing on the way pollution prevention has been promoted in the Netherlands during the past decade. This takes place in a collaborative setting. Regional authorities (provinces) play a central role in this by means of their so-called prevention teams. They develop projects in which consultancy agencies, research institutes, municipalities and trade associations collaborate in order to get individual firms active with pollution prevention. Overall, they conclude that most pollution prevention projects do result in better environmental performance by the participating companies, but for the majority it is a one-off experience. Not much is learned about the art of pollution prevention itself. The focus in the partnerships too often is on the actual performance of a company and finding technological improvements. Therefore they recommend that partners should look at pollution prevention projects as a learning process for the participating companies. Another problem is that current partnerships are not able to involve laggards through voluntary pollution prevention projects. To overcome these shortcomings De Bruijn and Hofman make some recommendations on how to improve the partnerships and the roles the various partners take on.

Part III of the book introduces cases in which a wide variety of actors tries to collaborate. These partnerships consist of companies, NGOs, governments, consumers, citizens, etc. The partnerships are often established in order to bring in broad or divergent views and are attempts for mutual learning. *Marilu Hastings* asserts that, through the process of establishing and managing collaborative relationships between firms and their stakeholders, firms do indeed develop new intangible resources, including new capabilities, enhanced reputation and social capital, which in turn lead to sustained competitive advantage. This argument is elaborated through the analysis of a case study of the stakeholder consultation process developed by Shell prior to commencing natural gas production in a remote area of

Peruvian rainforest. In a major break from its past standard operating procedure, the company adopted an unorthodox approach that incorporated extensive stakeholder participation to proactively identify and mitigate the environmental and social impacts of its activities and create net benefits to affected communities without providing direct cash payments. The stakeholder consultation process resulted in the development of new capabilities within Shell, the potential for Shell's enhanced reputation, and the development of new social capital that accrued to both parties. All of these intangible resources prove to be competitively valuable for the company.

These positive experiences with collaboration and consultation stand in contrast with the analysis that *Marie-France Turcotte and Basmah Ali* provide in their chapter. They examine the outcomes of a multistakeholders collaborative roundtable in the context of environmental problems. The case studied is the multi-stakeholder initiative ARET (Accelerated Reduction/ Elimination of Toxics), a roundtable organised at the national level by the Canadian Minister of the Environment in 1991 including representatives of governments, industries and NGOs. In this case the initial participants did achieve consensus on lists of targeted substances but did not reach consensus on what do about it. In the end it lead to the break-up of the initial multistakeholder group when the NGOs and the labour associations withdrew. Turcotte and Ali conclude that multi-stakeholder roundtables and voluntary approaches alone should not be expected to allow for self-regulation of the industry in the environmental domain. A government role in voluntary initiatives, such as ARET, is essential to earning and retaining public trust and ensuring continued effectiveness. Turcotte and Ali also state however that if multi-stakeholder roundtable should not be expected to replace regulations, however, they could be used in conjunction with regulations or other types of approaches to increase the capabilities for reaching environmental objectives.

Zinaida Fadeeva and Minna Halme focus on multi-stakeholder collaborative attempts in what they call the world's largest and fastest growing industry: tourism. In order to gain momentum to solve various environmental and socio-cultural questions, innovative actors in the field have begun to co-operate in the form of various partnerships like networks. Within such networks, cooperation bridges over business and public sector actors. Their study examines joint efforts of eight European networks of multiple actors toward sustainability. Their analysis reveals an interesting conclusion concerning the role of leadership: too strong a leadership may destroy the activity of other network members. While coordination and leadership secure efficient implementation of network actions and provide additional information sharing media, they may suppress the creation of new ideas, the natural re-formation of the network, and the emergence of sub-networks. They also conclude that the design and membership of the networks proved to be essential in dealing with complex and ambiguous tasks dealing with sustainable development. A too-narrow membership basis may limit the understanding of sustainable development within the network, whereas an overly large number of members may hinder implementation. They also find however that the composition of the membership does not determine the network's actions. A crucial success factor is the development of a common vision. The membership composition has to be related to this vision. Overall, it proved much more difficult in the cases they analysed to maintain networks than to establish them.

Jaco Quist, Ken Green, Klara Szita Toth and William Young focus in their chapter on how to achieve factor 10-20 environmental improvements in households. For this the involvement of a broad range of stakeholders is necessary, including consumers, companies, government, NGOs, research bodies and the media. Quist et al. describe and analyse a methodology they developed and used to create stakeholder involvement and collaboration. The methodology uses scenarios and workshops as a new method for involving stakeholders in sustainable development issues and generating new stakeholder partnerships and alliances. The aim was to involve experts from a range of stakeholder groups to develop ideas of sustainable shopping, cooking and eating in the year 2050. Despite the success of involving stakeholders and keeping them involved, the number of follow-up activities and stakeholder partnerships could be improved. However, these multi-actor participatory processes resulted in broad follow-up agendas. Each element of such agendas requires stakeholder co-operation and partnerships. A main issue the authors mention is how to fund these.

In a concluding chapter we take stock of the issues and experiences reported in the various chapters. First, an attempt is made to develop a somewhat more structured and coherent framework for assessing collaboration. Next, conclusions are drawn on the prospects and value of collaboration in the process of sustainable development.

REFERENCES

Bjerregaard, Ritt (1997) *New Possibilities for Industry*, speech to the seminar on the Development of EU Environmental Policies, Helsinki, 14 April.

CEC (1993) *Towards Sustainability: A European Community Programme of Policy and Action in relation to the Environment and Sustainable Development*, Luxembourg.

Clark, S. and S. Georg (1995) *From Greening to Sustaining: Transformational Challenges for the Firm, Summary Report of the third international conference of the Greening of Industry Network*, Copenhagen.

De Bruijn, T.J.N.M., P. Groenewegen and J. Grolin (1997) Global restructuring, A Place for Ecology, *Business Strategy and the Environment*, Vol. 6:173-184.

De Bruijn, T.J.N.M. and V. Norberg-Bohm (forthcomming) *Sharing responsibilities; Voluntary, collaborative and information-based approaches in environmental policies in the US and Europe.*

Fischer, Kurt and Johan Schot eds. (1993) *Environmental Strategies for Industry: International Perspectives on Research Needs and Policy Implications.* Washington DC: Island Press.

Gray, B. (1989) *Collaborating: Finding Common Ground for Multiparty Solutions,* San Francisco: Jossey-Bass.

Grolin, J. (1997) *Corporate legitimacy and risk society - Lessons from the Brent Spar conflict,* paper presented at the 14th Nordic Conference on Business Studies, 14-17 August, Bodø, Norway.

Hartman, C.L., P.S. Hofman and E.R. Stafford (1999) Partnerships: A Path to Sustainability, *Business Strategy and the Environment*, vol. 8, no. 5: 255-266.

Huber, J. (1995) *Nachhaltige Entwicklung. Strategien für eine ökologische und Sociale Erdpolitik,* Berlin.

International Human Dimensions Programme on Global Environmental Change (1999) *Industrial Transformation Science Plan,* IHDP Report No. 12, Bonn.

Lévêque, F. (1996) *Environmental Policy in Europe; Industry, Competition and the Policy Process,* Brookfield.

Long, F.J. and M.B. Arnold (1995) *The Power of Environmental Partnerships,* Fort Worth: Dryden.

Loudon, A.A. (1987) The chemical industry and the environment, *paper presented at the European Conference on Industry and Environmental Management,* Interlaken, Switzerland, October 2.

McDonough, W and M. Braungart (1998) The NEXT Industrial Revolution, *The Atlantic Monthly,* Vol. 282, No. 4: 82-92.

Roome, N. and E. Cahill (2001) *Sustainable production - challenges and objectives for EU research policy,* report of the expert group on competitive and sustainable production and related service industries in Europe in the period to 2020, EUR 19880 Research Directorate General, Final report.

Roome, N. (2001) *Policies and Conditions for Environmental Innovation and Management in Industry,* paper presented at the International Conference Towards Environmental Innovation Systems, Garmisch-Partenkirchen, September 27-29.

Schmidheiny, S. (1992) *Changing Course: A Global Business Perspective on Development and the Environment,* Cambridge.

Schot, J. (1997) Duurzame ontwikkeling, een uitdaging voor de industrie (Sustainable Development, A Challenge for Industry, in: K. Lulofs and G. Schrama, *Ketenbeheer (Chain Management),* Enschede.

Schot, J., E. Brand and K. Fischer (1997) The Greening of Industry for a Sustainable Future: Building an International Research Agenda, *Business Strategy and the Environment,* vol. 6, no. 3: 153-162.

Stigson, B. (2001) Converging toward a successful strategy, *European Eco-Efficiency Initiative Brief 3,* May 2001.

Von Weizsäcker, E.U., A.B. Hunter and L.H. Lovins (1997) *Factor Four. Doubling Wealth, Halving Resource Use, The new Report to the Club of Rome,* London.

World Commission on Environment and Development, *Our common future,* 1987.

Chapter 2

ENVIRONMENTAL COLLABORATION
Potential and Limits

CATHY L. HARTMAN

PETER S. HOFMAN

EDWIN R. STAFFORD

1. INTRODUCTION

Collaborative processes toward sustainability have been receiving increased interest among scholars, policy makers, business practitioners, and other environmental constituents (Hartman et al., 1999). Despite the emergence and acceptance of collaboration within and across a variety of sectors, including government, industry, and the environmental community, little critical research has contributed to our understanding of how effective and appropriate collaboration is as an alternative to traditional 'command and control,' protest, or confrontational approaches to environmental protection and sustainability (Harrison, 1999). In a review of collaborative partnerships amongst environmental non-government organisations (NGOs), businesses, and other entities, Murphy and Bendell (1997) claimed: "In most of the partnerships described ... almost no attempt was made to develop systems to evaluate the partnership's direct contributions to the achievement of environmental goals" (p. 229). In practice, some of the shortcomings resulting from collaboration to address environmental problems have generated harsh criticisms (Currah, 2000).

One commonly touted criticism is that collaboration tends to provide only incremental environmental progress rather than overt, discontinuous innovations or 'systems change' because collaboration does not allow for

T.J.N.M. de Bruijn and A. Tukker (eds.), Partnership and Leadership, 21–40.

sufficient changes in society's values and political structures necessary for sustainability (Levy, 1997). Collaboration typically occurs among those entities with the necessary resources and socio-political clout, such as government or industry, whose interests reinforce society's status quo and political structure. Collaboration tends to marginalise politically weaker groups, such as the poor, unemployed, less developed countries, future generations, or radical environmentalists (Levy, 1997). As a result, the distribution of environmental costs and benefits tends to be skewed toward the advantage of relatively powerful social groups and inhibits fundamental changes in social values and forces that contribute to environmental destruction. Such tenets as competitive individualism, affluence and consumption idealisation, utilitarian technical rationality over aesthetic or ethical values, and confidence in scientific progress and economic expansion are left unchallenged (Levy, 1997).

By contrast, others have pointed at the potential of collaboration as an 'action lever' in the transition towards sustainability (Murphy and Bendell, 1997). Environmental problems are commonly embedded within the competing interests of various society stakeholders, such as industry, government, labour unions, consumers, NGOs, etc., who commonly view one another as adversaries. Collaboration forges new relationships among these societal stakeholders that can facilitate the establishment of new social values compatible with sustainability. These new social values then provide the seeds for the kind of 'system change' deemed necessary for sustainable development. Initiating collaboration, however, appears to be a significant challenge, requiring either a mutually perceived crisis (Gray, 1989) or the threat of adverse regulation, protest, or negative publicity (e.g. Harrison, 1999) to bring parties together.

The opportunities and challenges of collaboration warrant careful consideration because the dominant theoretical paradigms shaping the debate on the transformation of environmental practices, discourses, and institutions all pose 'collaboration' as a necessary process for a more sustainable world. Drawing from research examining collaboration in government, industry, and environmental NGO sectors, this chapter contributes to the discourse about collaboration's potential for realising the kind of changes deemed necessary for sustainability by arguing that collaboration and conflict are ends of a continuum that need to be employed simultaneously. Because sustainability is not well defined, collaboration is needed to facilitate the constructive exploration of environmental stakeholder differences to establish collective understanding of the necessary path toward sustainability. The threat of conflict is needed frequently to encourage dialogue and participation.

Our discussion is in four parts. One, we discuss the concepts of sustainability and collaboration and the challenges collaboration poses in the pursuit of sustainability. Two, we overview the role collaboration plays within the dominant theoretical paradigms shaping the environmental transformation and sustainability debate - ecological modernisation, governance, and sociotechnical change theories. Three, we review emerging literature on trends and their reported outcomes across three key sectors engaged in collaboration - government, industry, and environmental NGOs. Last, based on this review and our synthesis of theory and research, we argue that, in practice, collaboration and more traditional regulatory and adversarial approaches can be conceptualised as ends of a continuum that need to be employed simultaneously to steer society onto a more sustainable path.

2. SUSTAINABILITY AND COLLABORATION

The concept of *sustainability* has come to occupy a permanent place on the public agenda (Hajer, 1995). On the one hand, the academic community, governments, businesses, and broader society recognise sustainability's importance for the future health and welfare of the planet and its inhabitants. On the other hand, an accepted definition of sustainability continues to be elusive. The most familiar definition comes from the 1987 Brundtland Commission Report entitled, *Our Common Future*, which describes sustainability as development that "meets the needs of the present without compromising the ability of future generations to meet their own needs" (1987:43). Since that publication, dozens of new interpretations and working definitions of sustainable development have appeared (e.g. Brooks, 1992), sparking debate over what sustainability really means and how to enact it (e.g. eco-efficiency) (Hartman et al., 1999).

Given the controversies surrounding the meaning of sustainability, Hart (1994) argues that the path to realising its principles will be long, arduous, and fraught with pitfalls. Therefore, sustainable development ought to be viewed more as a journey than a destination. As such, sustainability is as much a social transformation process as it is an environmental transformation process, requiring continuous capability building and management attention (Jones, 2000). Environmental challenges facing the world are complex where potential 'solutions' demand expertise from a variety of disciplines, government involvement from all levels, and international public support. Consequently, it is not possible for individual stakeholders working independently to find appropriate solutions, and collaboration offers one approach to foster transformation by creating a mechanism within which stakeholders can come to a shared understanding of and commitment to the actions and innovations necessary for sustainability.

David Murphy and Jem Bendell have recognised the necessity of a multistakeholder and process orientation with regard to sustainable development, and they have defined it as follows:

> "... a new organising principle, perhaps an emerging, positive myth, which has the potential to bring together diverse and often competing causes. The mythic quality of sustainable development lies in its capacity to clarify the earth's complexity and facilitate commitment to new collaborative models." (Murphy and Bendell, 1997:35)

Barbara Gray defines *collaboration* as "a process through which parties who see different aspects of a problem can constructively explore their differences and search for solutions that go beyond their own limited vision of what is possible" (Gray, 1989:5). Environmental stakeholders have unique interests and perspectives, particularly as they relate to the pursuit of sustainability; collaboration allows a more comprehensive appreciation of environmental challenges and what sustainability entails, encouraging expertise and resource exchanges. Collaboration implies interdependence among stakeholders, constructive handling of differences, joint ownership of decisions and collective responsibility of outcomes (Hartman et al., 1999).

Collaboration, however, poses significant challenges (Gray, 1989). It typically progresses from 'underorganised systems' of independent or adversarial stakeholders to a more tightly organised interrelationship characterised by co-ordinated decision-making. This co-ordination entails moving outside existing social structures, bringing together opposing values, exercising flexibility, and changing traditional independent and competitive mindsets among sectors of society. Historical and ideological barriers, institutional disincentives, social dynamics, risk perceptions, technical complexities and politics all present significant barriers inhibiting effective stakeholder participation necessary for sustainability (Gray, 1989). In practice, collaboration requires tolerance for compromise and a willingness to 'agree to disagree'. Thus, collaboration entails more than just economic or technological solutions to promote sustainability, but careful attention to relationships, decision-making fairness, and leadership (Chrislip and Larson, 1994).

David Chrislip and Carl Larson (1994) propose the concept of *collaborative leadership* as a necessary ingredient for partnerships. Hartman et al. have adapted this concept for sustainability by defining it as "the process to initiate, facilitate, and sustain collaborative initiatives among stakeholders for addressing environmental sustainability issues" (1999:257).

Collaborative leaders are not necessarily technical specialists or environmental experts. Rather, they are individuals and social entities who have the credibility and entrepreneurial initiative to bring the right individuals, organisations, and constituents together constructively to create visions, solve problems, and reach agreements. In sum, collaborative leaders are the catalysts for stakeholder collaboration.

The importance of collaboration as a mechanism for sustainability is featured in the three dominant theoretical paradigms shaping the environmental transformation debate – ecological modernisation theory, governance theory, and sociotechnical change theory. Next, we briefly overview these paradigms to highlight collaboration's potential role for sustainability.

3. COLLABORATION'S ROLE IN EMERGING THEORETICAL PARADIGMS

In developed countries, environmental politics came of age in the early 1970s, taking a predominantly legislative orientation to stem pollution emissions (Hajer, 1995). Environmental policy centred primarily on controlling air, water, and soil standards via quality targets or 'command-and-control' regulations on acceptable pollution emission levels from implicated industries. Pollution was perceived to be a problem that could be contained *ad hoc* with *ex post* remedial measures (Hajer, 1995). Typical abatement strategies under this regulatory regime were 'end-of-the-pipe technologies' such as filters on chimneys or drains, or water-processing plants.

By the 1980s, however, debate within political, industrial, and academic circles increasingly became critical of the traditional 'command-and-control' paradigm. Pollution was just one environmental problem, and other challenges, such as deforestation, nuclear waste, and ozone depletion, were recognised as endemic of the broader political, economic, and social institutions within society. The belief that reform and collective action across these sectors were necessary to address the environmental crisis became the foundation of what is widely recognised today as 'ecological modernisation' (Hajer, 1995; Mol, 2000).

Ecological modernisation theory is concerned with how contemporary industrialised societies deal with environmental crises (Mol and Sonnenfeld, 2000). It proposes that environmental reform of the modern organisation of production and consumption, and not necessarily a fundamental reorganisation of the core institutions of modern society such as the traditional environmental movement proposes, can shape a path towards sustainable development. Ecological modernisation theory "assumes that existing political, economic and social institutions can internalise the care for the environment" (Hajer, 1995:25). A basic tenet of ecological modernisation is "that the capitalist political economy needs conscious reconfiguring and far-sighted action so that economic development and environmental protection can proceed hand-in-hand and reinforce one another" (Dryzek, 1997:143). Ecological modernisation theory advances the idea that collaboration of key actors such as government, industry, reform-oriented environmentalists, and science, can generate win-win outcomes of economic development and environmental improvement.

Mol (2000) identifies three evolutionary stages in the conceptual development of ecological modernisation. In the early 1980s, Joseph Huber championed technological innovation, especially in industrial production, as a means of improving environmental quality. His work took a critical view of the bureaucratic nature of environmental policy, favouring market mechanisms. In the late 1980s, the debate shifted away from technology to focus more on government and market dynamics in ecological transformation. Attention was given to institutional and cultural issues related to environmental protection, and studies emerged on industrial production and consumption processes. This research centred largely on Western European countries. From the mid-1990s onward, the literature has showed increased attention to globalisation and international dynamics on environmental improvement, particularly in the developing world and the USA.

Hajer (1995) asserts that ecological modernisation centres on three principle beliefs: (1) environmental degradation can be evaluated by combining cost-benefit analysis with science; (2) with appropriate design and structure, economic development and environmental protection can go hand-in-hand; and (3) collective action is necessary to achieve environmental protection. In sum, the premise of this school of thought centres on a belief that collective action and the need for socio-technical and economic reforms rather than command-and-control regulations can address and resolve environmental problems.

Governance theory has in the last decade analysed new forms of steering alternatives to hierarchical control models, especially more on co-operation focussed models in which governmental, non-governmental and private actors participate in mixed public and private networks (Mayntz, 1998). Environmental governance models emerge as alternatives to traditional forms of direct regulation because it becomes evident that conventional command-and-control models are not able to cope with 'wicked' environmental problems. Governance points to the interdependence among actors, their interactions, and the rules associated with them. It also points to the importance of collaboration as "no single actor, public or private, has all knowledge and information required to solve complex, dynamic and diversified problems" (Kooiman, 1993:4). Governance theory recognises the increasing importance of civil society, for example, in co-setting the agenda in processes of local and global environmental change.

Governance thus diverges from paradigms where either hierarchy or the market is the dominant organisational form, and emphasises the role of networks in organising relations between actors. Dominant processes of governing in networks concern negotiation, accommodation, concertation, co-operation and alliance formation, rather than the traditional processes of coercion and command and control. In environmental policy, the increasing interest in the concept of governance has shaped the emergence of co-operative environmental strategies between public and private actors. Examples are voluntary agreements between industry and government (Glasbergen, 1998) and the stimulation of environmental management in firms through the use of policy networks based on consensual steering models (De Bruijn and Lulofs, 2000). The involvement of partners such as national governments, business, and consumers is also seen as crucial in European environmental policy. The fifth Environmental Action Plan of the European Union for instance stresses the importance of shared responsibilities among governments, business and the general public out of the understanding that the ultimate goal of sustainable development can only be achieved by relevant actors working together in partnership (CEC, 1993:113).

Whereas the concept of governance has mostly been associated with developments in the public sector, it also has impacted the corporate sector. The concept of corporate governance captures the increasing role of a variety of stakeholders, and deals with the way companies manage stakeholder relations. It also acknowledges that demands for better environmental performance, and increased accountability and transparency increasingly emerge from civil society rather than the state, especially because companies' activities are increasingly global in nature and difficult to capture in specific national boundaries (Bendell, 2000). Companies need a social licence to operate as well as the traditional regulatory licence (Warhurst, 2001), and research on governance focuses on corporate approaches in balancing social and environmental responsibility with profitability in interaction with its stakeholders (Halal, 2001).

The concept of governance refers to the interaction among actors in state, civil society, and industry, their interrelatedness at both local, national and global levels, and has given rise to various co-operative strategies for environmental management and sustainable development. The strengthening of civil society through stakeholder partnerships has been a driving force for curtailing industrial pollution, conserving nature, and introducing more sustainable corporate practices (Shrivastava, 1995; Guha and Martinez-Alier 1997; Hofman, 1998; Lemos, 1998; Stafford et al., 2000). Academic research in this field has sharply increased in the last decade, but there is still need for theory building as to how and under what conditions these strategies contribute to sustainable development.

Sociotechnical change theory advocates the integrated study of society and technology and has provided insights on the way technology evolves in society and social shaping of technology occurs (Bijker et al., 1987; Williams and Edge, 1996; Rip and Kemp, 1998). It considers technological change not as a rigid, categorised, process but as a multidirectional flux that involves constant negotiation and renegotiation among and between groups shaping the technology (Pinch and Bijker, 1987). Technology is shaped by social, economic, cultural and political forces and in the same process technology shapes human relations and societies (Rip and Kemp, 1998). This coevolution is often path dependent in the sense that configurations of technology entrenched in social processes, consumption patterns, and lifestyles are difficult to reverse. The concept of path dependence is useful in order to explain why, despite clear intensification of environmental pressures, policies have often not been able to foster fundamental changes in production and consumption. Environmental measures, technologies and policies initiate from the accumulated technologies and competencies that have brought forth technological systems entrenched in various institutional structures and embedded in society. Here sociotechnical change theory also shares common ground with new institutional theory (March and Olsen, 1989; Powell and DiMaggio, 1991; Guy Peters, 1999) that points to the importance of institutions (both formal and informal rules) as sources for path dependence. Sociotechnical change theorists particularly explore how technologies and artefacts are both products and sources of these path dependencies. Both technological and regulatory development can become locked in a path of gradual improvement of dominant systems and are, consequently, unable to bring forth fundamental systems change.

Sociotechnical change theory points to collaboration as a way out of this deadlock. Social networks are key both in the stabilisation of present technologies and, potentially, in the creation of new ones (Weaver et al., 2000). Path dependence can occur due to routinised behaviour and R&D trajectories that become fixed around dominant guiding principles. As problem definitions become shared within a network a powerful constituency is developed that perpetuates its technology. Based on experiences of a program on sustainable technology development Weaver et al. (2000) argue that new directions for R&D might be found by creating new cross-sectoral networks around innovation challenges, and by helping network members to redefine innovation challenges in new terms. From a similar perspective Schot (2001) proposes to broaden the design process of technologies by bringing together all interested parties early on and throughout the design process. In this way potential for mutual learning regarding technological options, user preferences, institutional aspects and environmental impacts can be fully exploited and facilitate embedding of new (sustainable) needs and values in material (technology) realities (Schot, 2001).

In sum, sociotechnical change theorists aim to increase understanding of the dynamics and patterns of coevolutionary processes that inform where opportunities exist to trigger new actor linkages and alignments which can enable the creation of new transformational paths (Kemp, Rip and Schot, 2001). The notion of new actor linkages for sustainability parallels governance theory's increasing emphasis on the role of public and private sector networks and alliances and ecological modernisation's concept of stakeholder co-ordination to pursue win-win outcomes for economic development and environmental improvement. On balance, the dominant theoretical paradigms shaping the environmental transformation debate all point to collaboration as a cornerstone to sustainability. Next, we review how environmental collaboration has developed within the government, industry, and environmental NGO sectors. Though not an exhaustive review, it highlights specific cases to articulate the environmental progress as well as shortcomings collaboration has experienced in practice. Our review illuminates some guideposts for understanding the potential for environmental collaboration and the need for traditional regulatory and confrontation approaches to encourage greener and cleaner industrial practices.

4. ENVIRONMENTAL COLLABORATION IN PRACTICE

Government

The global nature of environmental problems, fiscal constraints, and an informed public requires a new anticipatory, systems-based regulatory strategy to build stakeholder consensus and commitment to continuous environmental improvement. New modes of governance have arisen which assign a key role to collaboration among government, mediating NGOs, and private interests (Glasbergen, 1998; Midttun, 1998). Basic features of these types of management are collaboration of various societal groups debating and negotiating environmental problems to reach a common understanding regarding the parameters of environmental issues and appropriate ways to address problems (Meadowcroft, 1998).

Ten Brink et al. (1998), for example, provide an assessment of the effectiveness of environmental agreements in European countries. Environmental agreements are defined by these authors as "those commitments undertaken by firms and sectors, which are the result of negotiation with public authorities and/or explicitly recognised by the authorities" (Ten Brink et al., 1998:2). They conclude that, in most cases, environmental agreements have contributed to improved performance. Moreover, environmental agreements, which complement existing regulation, are generally viewed by government to be at least as effective as traditional regulation and enforcement. Overall success of these kinds of partnerships depends, however, on the appropriateness of institutional structures (e.g., closely-linked, homogeneous industries), credible threats of sanctions, and incentives. While environmental agreements seem to imply a move away from traditional regulation, opinions vary as to whether consensual approaches have the potential to realise sustainable development. It is argued that because they are building on existing socio-economic structures, new consensual approaches favour incremental rather than significant socio-political and industrial changes (Gibbs, 1998; Blowers, 1998). Early findings of evaluations of environmental agreements tend to confirm this proposition (Ten Brink et al., 1998; Glasbergen, 1998).

A comparison of research on governance issues highlights the large gap in experience between countries in the North and South. Routes for improving environmental performance of industries through the introduction of environmental management systems and/or pollution prevention are at a preliminary stage for much of the developing world, as policies have only recently deviated from traditional command and control paths (Wehrmeyer and Mulugetta, 1999). In their study of small-scale industries in India, Clancy and Lakmakker (1998) report a need for strengthening of the administrative capacity for pollution control. While they identify various opportunities for improved environmental performance in their case study of the textile dyeing industry, they conclude that environmental change is either non-existent or slow due to lack of technological and managerial capabilities, fierce competition, and the absence of enforcement of environmental regulation. This research suggests the greening of industry and the role of government in industrialising and less industrialised countries are very different from that of industrialised countries. This questions also the relevance and applicability of theory emerging from industrialised settings to environmental issues in developing countries.

Sustainable development as a global issue includes an analysis of new linkages among actors as producers, consumers, workers, and citizens in both developed and developing countries. Because many of these linkages cross geographical borders, the role of national policies in addressing these issues is limited. What matters most are self-regulatory 'codes of conduct' by industry itself and especially actions by NGOs.

Although interest in co-operative approaches has increased in the last decade there is still considerable uncertainty regarding their effectiveness in different circumstances. This has been the subject of various studies from both North American (e.g. Weber, 1998; Paton, 1999; Harrison, 1999) and European scholars (e.g. Glasbergen, 1998, 1999; Carraro and Lévêque, 1999; Mol, Lauber and Liefferink, 2000). Some of the potential benefits of co-operative approaches are that they lower transaction costs relative to direct regulation (Weber, 1998) and that they could facilitate the choice of better solutions to environmental problems by taking advantage of business expertise (Harrison, 1999). Based on the Dutch experiences with environmental covenants and long term agreements on energy efficiency improvement Glasbergen (1999) stresses the importance of learning in the process. He defines the covenanting process as a structured dialogue between public and private parties that exchanges data, interpretations and perspectives, and gain understanding of mutual dependencies, various problem perspectives, and alternative solutions. Joint decision making and acceptance of shared responsibility are then critical factors in this form of governance (1999:57). However, they need to take place in a framework where clear targets are set and for which results are monitored and accounted.

In an overview of environmental agreements, Harrison (1999) poses a provocative issue with regard to governments using coercive versus co-operative approaches with businesses. Harrison notes that the typical incentive for businesses to consider co-operative environmental agreements centres on avoiding burdensome regulations. "Therein lies the paradox of a co-operative voluntary approach: A coercive government is often a necessary prerequisite for co-operative agreements" (1999:59). That is, collaborative environmental agreements may be feasible only when the government poses a legitimate threat to regulate or fine a business. To date, evaluation of the environmental and economic benefits of governmental collaborative agreements has yet to receive much attention, in part because of unclear targets and insufficient monitoring and verification schemes. Moreover, environmental and economic benefits may vary across sectors and countries. Such benefits may depend on how governments strike the right balance between demonstrating commitment to negotiate and maintaining a credible threat of regulation. In addition, a fixed composition of actors within the process can also create lock-in to a path of incremental change. These areas remain challenges for further research.

Industry

While issues of environmental management are still central among the debates and research on corporate environmental practice, current studies reflect a growing recognition of the embeddedness of processes of industrial transformation in societal change (Hartman et al., 1999). Earlier research on environmental management issues focused primarily on the identification of internal and external forces driving corporate environmental management, but a paradigm shift is taking place that views integration of environmental considerations in corporate management as a 'collaborative learning process' (Clarke and Roome, 1999; Bendell, 2000; Halal, 2001). Implementation of values associated with sustainable development is therefore a process in which collaboration is indispensable. Clarke and Roome (1999; in this volume) argue that effective environmental management and sustainable development involve inclusive networks for learning and action, requiring companies to use networks of stakeholders as a means to confirm and validate their approaches to environmental problems. Not only does their 'learning-action network' concept require multi-stakeholder consultations and collaboration, but it also implies a high degree of openness and transparency concerning corporate operations and management or collaborative leadership. This participative process is a precondition for mutual learning and transformation.

Hart (1995) identifies three strategic capabilities that form the core of the firm's environmental management contributing to sustained competitive advantage: pollution prevention, product stewardship, and sustainable development[1]. In order to acquire these capabilities, companies increasingly need to employ complex collaborative strategies. Pollution prevention involves optimal resource uses and minimisation of emissions along with tacit skill development. Product stewardship centres on the minimisation of life-cycle costs of products and involves socially complex skills and involvement (and collaboration) of external stakeholders. Sustainable development requires minimisation of the environmental burden of the firm's activities, development, and product systems. It involves dialogue, consultation, and collaboration with stakeholders regarding company performance and company values at both local and global levels. Hart proposes that those companies who develop and establish a shared vision of the future can acquire the capability of sustainable development. The strategic capabilities are interconnected in the sense that acquiring skills for pollution prevention is a prerequisite for product stewardship, and demonstrating capabilities for product stewardship will pre-empt capabilities for sustainable development (Hart, 1995). While Hart's framework is predominantly built on the experience of larger companies, Medina-Munoz and Garcia-Falcon (1998) apply the framework to small- and medium-size enterprises (SMEs). They assert that SMEs can gain competitive advantages by deploying a sustainability strategy at the local level, especially by linking

[1]See also chapter 11 of this book in which Hart's model and the role of pollution prevention is further discussed.

with the sociocultural aspects of communities, such as conserving the local cultural and environmental heritage. Given that SMEs are frequently already forced to cooperate with other organisations to achieve economies of scale, make investment decisions locally, and their organisational cultures are expected to fit to community values and attitudes, they represent credible partners in collaboration.

There is evidence that in some cases the voluntary aspect of co-operative programs creates opportunity for external manipulation of stakeholders or 'green washing' rather than internal change and commitment to the desired social agenda. In their study of Responsible Care, the oldest industry code of environmental, health, and safety (EHS) practice, Howard et al. (2000) concluded that this code elicited the greatest and most uniform response from companies in areas that were open to public scrutiny – community relations and distribution. Additionally, companies who responded by making changes in internal as well as external programs, were those who were committed to EHS improvements prior to the enactment of Responsible Care. Their findings suggest that companies respond to institutional pressures strategically reflecting how members (1) believe outsiders view their organisation and (2) understand their organisation's central beliefs and values. In the case of Responsible Care, the opportunity for discretion and weak sanctions for non-compliance resulted in a compliance level that was consistent with an organisation's pre-code level of commitment. Consequently, the industry did not move beyond the level that would have been reached without the code, and it allowed uncommitted firms the opportunity to hide behind the image created by the mere existence of the Responsible Care initiative. In short, these findings parallel Harrison's (1999) assertion that without a coercive threat of sanctions from either government or other social entities, voluntary collaborative agreements may be ineffectual.

Environmental Non-Government Organisations (NGOs).

Often called the 'third sector' outside of government and industry (Hudock, 1999:1), NGOs are characterised typically by their non-profit status and value-based orientation. In the environmental arena, NGOs link local and international levels of politics to advance their self-adopted moral missions. In his report to the 53rd Session of the United Nations General Assembly, Secretary-General Kofi Annan (1998) described NGOs as: "no longer seen only as disseminators of information, but as shapers of policy and indispensable bridges between the general public and the intergovernmental processes".

The culture of the environmental community is shifting from one of protest to practical solutions (Murphy and Bendell, 1997). Increasingly, environmental NGOs have become interested in 'working the system'

(Milne et al., 1996:203), advocating market, social, and public policy reforms to encourage environmentally sensitive corporate practices over 'command-and-control' regulatory mechanisms that typically pit business interests against environmental concerns (Lober, 1997). During the 1990s, collaborating directly with industry has emerged as a pivotal, if controversial, strategy for furthering sustainable business practices. Stafford and Hartman (1996) have termed these unique initiatives as 'green alliances', defined as collaborative partnerships between environmental NGOs and business (and sometimes other stakeholders) to pursue mutually beneficial ecological goals. Savvy environmental NGOs typically entice businesses into collaboration by positioning initiatives as avenues for positive publicity, increased credibility, total quality management, resource reductions, cost-savings, and new technologies (Hartman and Stafford, 1997). Thus, green alliances seek the 'common ground' among ecological, social, and commercial interests necessary for sustainability.

In green alliances, environmental NGOs may assist corporate environmental initiatives in two ways: one, environmental NGOs can provide corporations with ecological, scientific, and legal expertise (Milne et al., 1996), and two, environmental NGOs can leverage and broker corporate linkages with other diverse stakeholders, such as the media, consumers, or government, whose support may be critical to credibly position a firm's environmental initiatives (Stafford et al., 2000). These benefits are exemplified by Greenpeace Germany's alliance with Foron, a small east German appliance manufacturer on the verge of bankruptcy, to market 'greenfreeze' refrigerators in 1992 (Porter and Van der Linde, 1995). Greenpeace provided Foron with greenfreeze, an environmentally benign propane-butane coolant technology, as an alternative to the then-prevalent cooling agents - hydrofluorocarbons and hydrochlorofluorocarbons - proposed as replacements for chlorofluorocarbons. To facilitate market acceptance, Greenpeace lobbied for government certification of the refrigerant, procured development funding, galvanised media attention, and stimulated consumer demand with an advertising campaign for Foron's refrigerators. Market and socio-political acceptance of greenfreeze ultimately encouraged most of Germany's refrigerator industry to switch to the technology by 1994 (Stafford et al., 2000). Today, greenfreeze is the technology standard throughout Europe, and it has made significant inroads in the developing world.

While the environmentalist-business exchanges inherent in green alliances may facilitate the corporate objectives of strategic planning, differentiation, cost savings, and regulatory flexibility (Lober, 1997), such partnerships face the challenges of integrating market and environmental objectives, which typically result in compromises (Stafford and Hartman, 1996). As a result, NGOs risk their public credibility when they engage in industry partnerships (Currah, 2000; Long and Arnold, 1995; Stafford and

Hartman, 2000). This is exemplified by the case of Pollution Probe's endorsement of a private label product line called 'GREEN' from Canada's largest food retailer, Loblaws, in 1989 (Westley and Vredenburg, 1991). Pollution Probe intended that its certification would educate consumers, shift market demand toward environmentally preferable products, and ultimately influence industry to become more environmentally sensitive. The collaborative arrangement, however, proved problematic. For competitive reasons, the endorsement was negotiated in secrecy and provided the NGO with a 1% royalty on sales to cover the NGO's research and certification expenditures. When the endorsement was made public, many Pollution Probe staff members and scientists who were not briefed beforehand resigned in protest over what they perceived as an inappropriate relationship. Greenpeace was suspicious of the certification and analysed Loblaws' GREEN fertiliser, finding traces of toxins. Greenpeace publicly criticised Pollution Probe's ambiguous certification criteria and accused the partnership as simply exploitation to sell products and raise funds for the NGO. The negative publicity ultimately forced Pollution Probe to end the endorsement, and it became a weaker, less credible environmental force (Murphy and Bendell, 1997).

Currah (2000) argues that the trend of green alliances and the resulting compromised consequences of collaboration could jeopardise the NGO sector's "original mandate to address the plight, political demands, and individual aspirations of particular constituencies within civil society" (p. 348). Currah asserts that as NGOs frame and adapt their environmental missions to those of corporations, they are destined to become 'quasi-corporations' themselves, losing their public legitimacy as autonomous advocates for the environment. "Established NGOs are in danger of being superseded by newer organisations that will be more radical - yet more representative [of public interests] - if they do not ensure that the objectives of their alliances are grounded in their constituents' best interests" (p. 348). In sum, the emergence of market-oriented and collaborative campaigns amongst environmental NGOs poses some serious pitfalls, particularly with respect to NGO credibility.

Greenpeace sees confrontation and partnership as the ends of the same continuum, and their campaign to diffuse greenfreeze world-wide has employed a savvy blend of traditional adversarial tactics and collaboration (Stafford and Hartman, 2001). Understanding when to use either confrontation or collaboration becomes key to maintaining NGO credibility amongst prospective business foes and allies as well as general society. The idea that NGOs must maintain their adversarial orientation in society parallel's Harrison's (1999) observed paradox of collaboration between government and business; like government, NGOs must pose a threat to convince business of their legitimacy as entities worthy of negotiation and partnership.

This section has attempted to provide an overview of emerging research examining the practice of collaboration and its recognised opportunities and limitations as evidenced from the government, industry, and environmental NGOs sectors. Next, we incorporate these perspectives to conclude our argument that the path toward sustainability warrants the synergy of both collaboration and more traditional adversarial approaches.

5. COLLABORATION AND CONFLICT: A SYNERGY?

Evidence suggests that several characteristics of collaboration make it highly suitable for addressing environmental problems and steering society toward sustainability. First, environmental problems are complex and actions must be taken without benefit of conclusive evidence. Collaboration offers the opportunity to bring together experts from a variety of different disciplines and arenas to fashion solutions that can go beyond the limited perspectives of individual stakeholders. Such networks can also be critical for stakeholder learning (Clarke and Roome, 1999). Second, environmental solutions often require a cadre of economic, political, social, and marketing resources. Collaboration creates a stakeholder network for the marshalling of necessary resources. Third, environmental solutions typically bring new stakeholders to the table who must work together in novel ways. As opposed to traditional command and control or adversarial approaches, collaboration can work outside the formal political, social, and economic structures in society with the flexibility to build stakeholder networks relevant to the problem at hand. In short, collaboration is essential to sustainability.

In practice, collaboration has not been a panacea for environmental transformation. The cases reviewed here indicate that collaboration's contributions can be incremental or limited if a threat of regulation, sanction, or protest is absent. The challenge is encouraging relevant stakeholders to look beyond their own limited, short-term self-interests and recognise the broader, long-term benefits inherent in dialogue and collaboration, and this may only occur when stakeholders mutually perceive a crisis (Gray, 1989) or the threat of coercion (Harrison, 1999).

Defining sustainability as a social process suggests that environmental issues and solutions can only be defined through dialogue among stakeholders. Environmental reality is socially produced; only when stakeholders become aware that their perceptions about reality are not, in fact, reality will an impetus for change exist. In cases where commercial, social, and political institutions are resistant to change, traditional adversarial approaches, such as the threat of regulation or confrontation, may be necessary to initiate the process. On balance, rather than characterising traditional adversarial approaches and collaboration as either/or paths for sustainability, the two may actually go hand in hand in a synergistic fashion.

REFERENCES

Annan, K. (1998) *Strengthening of the UN System*, Report to the 53rd Session of the UN General Assembly, Document A/53/170.

Bendell, J. ed. (2000) *Terms for Endearment, Business, NGOs and Sustainable Development*, Sheffield: Greenleaf Publishing.

Bijker, W.E, T.P. Hughes and T. Pinch eds. (1987) *The Social Construction of Technological Systems*, Cambridge: MIT Press.

Blowers, A. (1998) Power, Participation and Partnerships: The Limits of Co-operative Environmental Management, in *Co-operative Environmental Governance, Public-Private Agreements as a Policy Strategy* P. Glasbergen ed. 229-249, Dordrecht: Kluwer Academic Publishers.

Brooks, D.B. (1992) The Challenge of Sustainability: Is Integrating Environment and Economics Enough? *Policy Sciences* 26: 401-8.

Brundtland, G-H. (1987) *Our Common Future* – report of the World Commission on Environment and Development. Oxford: Oxford University Press.

Carraro, C. and F. Lévêque eds. (1999) *Voluntary Approaches in Environmental Policy*, Dordrecht: Kluwer Academic Publishers.

CEC (1993) *Towards Sustainability: A European Community Programme of Policy and Action in relation to the Environment and Sustainable Development*, Luxembourg.

Chrislip, D.D. and C.E. Larson (1994) *Collaborative Leadership: How Citizens and Civic Leaders Can Make a Difference*, San Francisco: Jossey-Bass.

Clancy, J. and J. Lakmakker J. (1998) *Improving the Environmental Performance of Small-Scaled Industries in the South: A Case Study of the Dyestuffs Industry in India*, Paper presented at the Greening of Industry Network Conference, Rome, Italy.

Clarke, S. and N.J. Roome (1999) Sustainable Business: Learning-Action Networks as Organisational Assets, *Business Strategy and the Environment*, 8 (5): 296-310.

Currah, K.R. (2000) Activating the Self-Correcting Mechanism of Civil Society: A Critical Analysis of TNC-NGO Relations, *International Journal of Public-Private Partnerships*, 2 (3): 347-365.

De Bruijn, T.J.N.M. and K.R.D. Lulofs (2000) Driving SMEs towards Environmental Management: Policy Implementation in Networks, in *Small and Medium-sized Enterprises and the Environment: Business Imperatives* R. Hillary ed. 263-274, Sheffield: Greenleaf Publishing.

Dryzek, J.S. (1997) *The Politics of the Earth: Environmental Discourses*, Oxford: Oxford University Press.

Gibbs, D. (1998) *Ecological Modernisation: A Basis for Regional Development?* Paper presented at the Seventh Greening of Industry Network Conference, Rome, Italy.

Glasbergen, P. ed. (1998) *Co-operative Environmental Governance, Public-Private Agreements as a Policy Strategy*, Dordrecht: Kluwer Academic Publishers.

Glasbergen, P. (1999) Tailor-Made Environmental Governance: On the Relevance of the Covenanting Process, *European Environment*, 9: 49-58.

Gray, B. (1989) *Collaborating: Finding Common Ground for Multiparty Solutions*, San Francisco: Jossey-Bass.

Guha, R and J. Martinez-Alier (1997) *Varieties of Environmentalism: Essays North and South*, London: Earthscan.

Guy Peters, B. (1999), *Institutional Theory in Political Science; The 'New Institutionalism'*, New York: Pinter.

Hajer, M.A. (1995) *The Politics of Environmental Discourse, Ecological Modernisation and the Policy Process*, Oxford: Oxford University Press.

Halal, W.E. (2001) The Collaborative Enterprise, A Stakeholder Model Uniting Profitability and Responsibility, *Journal of Corporate Citizenship*, 2 (Summer): 27-42.

Harrison, K. (1999) Talking with the Donkey: Co-operative Approaches to Environmental Protection, *Journal of Industrial Ecology*, 2 (3): 51-72.

Hart, S.L. (1994) How Green Products Might Sustain the World, *Journal of the Northwest Environment*, 10: 4-14.

Hart, S.L (1995) A Natural-Resource-Base View of the Firm, *Academy of Management Review*, 20 (4): 986-1014.

Hartman, C.L. and E.R. Stafford (1997) Green Alliances: Building New Business with Environmental Groups, *Long Range Planning*, 30 (2): 184-196.

Hartman, C.L., P.S. Hofman and E.R. Stafford (1999) Partnerships: A Path to Sustainability, *Business Strategy and the Environment*, 8 (5): 255-266.

Hofman, P.S. (1998) Participation in Southeast Asian Pollution Control Policies, in *Participation and the Quality of Environmental Decision Making* F.H.J.M. Coenen, D. Huitema and L.J. O'Toole eds. 287-305, Dordrecht: Kluwer Academic Publishers.

Howard, J., J. Nash, and J. Ehrenfeld (2000) Standard or Smokescreen? Implementation of a Voluntary Environmental Code, *California Management Review*, 42, 63-82.

Hudock, A.C. (1999) *NGOs and Civil Society: Democracy by Proxy?* Cambridge: Polity Press.

Jones, D.R. (2000) Leadership Strategies for Sustainable Development: A Case Study of Suma Wholefoods, *Business Strategy and the Environment*, 9 (6): 378-389.

Kemp, R., A. Rip and J. Schot (2001) Constructing Transition Paths Through the Management of Niches, in *Path Dependence and Creation* R. Garud and P. Karnoe eds. Chapter 6, 269-299, Mahwah: Lawrence Erlbaum Associates.

Kooiman, J. ed. (1993) *Modern Governance: New Government-Societylinteractions*, London: Sage.

Lemos, M.C. (1998) Popular Participation and Pollution Control in Brazil, in *Participation and the Quality of Environmental Decision Making* F.H.J.M. Coenen, D. Huitema and L.J. O'Toole eds. 267-285, Dordrecht: Kluwer Academic Publishers.

Levy, D.L. (1997) Environmental Management as Political Sustainability, *Organisation & Environment*, 10 (2): 126-147.

Lober, D.J. (1997) Explaining the Formation of Business-Environmentalist Collaborations: Collaborative Windows and the Paper Task Force, *Policy Sciences*, 30 (1): 1-24.

Long, F.J. and M.B. Arnold (1995) *The Power of Environmental Partnerships*, Fort Worth: Dryden.

March, James G. and Johan P. Olsen (1989) *Rediscovering Institutions; The Organisational Basis of Politics*, The Free Press.

Mayntz, R. (1998) New Challenges to Governance Theory, *Jean Monnet Chair Papers*, No. 50, Florence: European University Institute.

Meadowcroft, J. (1998) Co-operative Management Regimes: A Way Forward? in *Co-operative Environmental Governance, Public-Private Agreements as a Policy Strategy* P. Glasbergen ed. 21-42, Dordrecht: Kluwer Academic Publishers.

Medina-Munoz, D., and J.M. Garcia-Falcon (1998) *Sustainability as a Major Source of Competitive Advantage for Small and Medium Sized Enterprises*, Paper presented at the Seventh Greening of Industry Network Conference, Rome, Italy.

Midttun, A. (1998) *The Weakness of Strong Governance and the Strength of Soft Regulation: Environmental Governance in Post-Modern Form*, Paper presented at the Seventh Greening of Industry Network Conference, Rome, Italy.

Milne, G.R., E.S. Iyer and S. Gooding-Williams (1996) Environmental Organisation Alliance Relationships Within and Across Nonprofit, Business, and Government Sectors, *Journal of Public Policy and Marketing*, 15 (Fall): 203-215.

Mol, A.P.J. (2000) The Environmental Movement in an Era of Ecological Modernisation, *Geoforum*, 31 (1): 45-56.

Mol, A.P.J. and D.A. Sonnenfeld (2000) Ecological Modernisation Around the World, An Introduction, *Environmental Politics*, Vol. 9, 1, 3-14.

Mol, A.P.J., V. Lauber and D. Liefferink eds. (2000) *The Voluntary Approach to Environmental Policy*, Oxford: Oxford University Press.

Murphy, D.F. and J. Bendell (1997) *In the Company of Partners: Business, Environmental Groups, and Sustainable Development post-Rio*, Bristol: The Policy Press.

Paton, B. (1999) Voluntary Environmental Initiatives and Sustainable Industry, *Business Strategy and the Environment*, Vol. 9, 5, 328-338.

Pinch, T.J. and W.E. Bijker (1987) The Social Construction of Facts and Artifacts: Or How the Sociology of Science and the Sociology of Technology Might Benefit Each Other, in *The Social Construction of Technological Systems* W.E. Bijker, T.P. Hughes and T. Pinch eds. 17-50, Cambridge: MIT Press.

Porter, M.E. and C. Van der Linde (1995) Green and Competitive: Ending the Stalemate. *Harvard Business Review* 73 (5): 120-134.

Powell, W.W. and P.J. DiMaggio eds. (1991) *The New Institutionalism in Organisational Analysis*, Chicago: University of Chicago Press.

Rip, A. and R. Kemp (1998) Technologic Change, in *Human Choice and Climate Change* S. Rayner and E.L. Malone, Volume Two, Resources and Technology, Chapter 6, 327-399, Columbus: Batelle Press.

Schot, J. (2001) Towards New Forms of Participatory Technology Development, *Technology Analysis and Strategic Management*, Vol. 13, No. 1: 39-52.

Shrivastava, P. (1995) Democratic Control of Technological Risks in Developing Countries, *Ecological Economics* 14: 195-208.

Stafford, E.R. and C.L. Hartman (1996) Green Alliances: Strategic Relations Between Businesses and Environmental Groups, *Business Horizons*, 39 (March-April): 50-59.

Stafford, E.R., M.J. Polonsky and C.L. Hartman (2000) Environmental NGO-Business Collaboration and Strategic Bridging: A Case Analysis of the Greenpeace-Foron Alliance, *Business Strategy and the Environment*, 9 (2): 122-135.

Stafford, E.R. and C.L. Hartman (2001) Greenpeace's 'Greenfreeze Campaign': Hurdling Competitive Forces in the Diffusion of Environmental Technology Innovation, in *Ahead of the Curve: Cases of Innovation in Environmental Management*, K. Green, P. Groenewegen, and P.S. Hofman eds. 107-131, Dordrecht: Kluwer Academic Publishers.

Ten Brink, P., V. Pollard and J. Medhurst (1998) *Environmental Agreements: Environmental Effectiveness and Potential for Future Use,* Paper presented at the Seventh Greening of Industry Network Conference, Rome, Italy.

Warhurst, A. (2001) Corporate Citizenship and Corporate Social Investment, Drivers of Tri-Sector Partnerships, *Journal of Corporate Citizenship*, 1, Spring, 57-73.

Weaver, P., L. Jansen, G. van Grootveld, E. van Spiegel and P. Vergragt (2000) *Sustainable Technology Development*, Sheffield: Greenleaf Publishing.

Weber, E.P. (1998) *Pluralism by the Rules, Conflict and Co-operation in Environmental Regulation*, Washington DC: Georgetown University Press.

Wehrmeyer, W. and Y. Mulugetta eds. (1999) *Growing Pains: Environmental Management in Developing Countries*, Sheffield: Greenleaf Publishing.

Westley, F. and H. Vredenburg (1991) Strategic Bridging: The Collaboration Between Environmentalists and Business in the Marketing of Green Products, *Journal of Applied Behavioral Sciences*, 27 (March): 65-90.

Williams, R. and D. Edge (1996) The Social Shaping of Technology, *Research Policy*, 25, 865-899.

Chapter 3

IN SEARCH OF THE 'WIN-WIN'

Multistakeholder Environmental Partnerships and the Pursuit of Sustainability

ERIC C. PONCELET

1. INTRODUCTION

Throughout the industrialised West, the conjunction of receding welfare states (Buttel, 1992), prominent market liberalisation ideologies (Gill, 1995), mounting dissatisfaction with conventional regulatory regimes (Collier, 1998; Golub, 1998), and persistent frustration over gridlocked decision-making processes (Anderson, 1997) has resulted in the increased promotion of voluntary, collaborative approaches in the environmental arena. Multistakeholder environmental partnerships have emerged as an important manifestation of this trend, making significant contributions in the areas of environmental planning, management, and policymaking (Gray, 1989; Long and Arnold, 1995; Wondolleck and Yaffee, 2000) as well as in the domain of participatory democracy (Fiorino, 1996; Lafferty and Meadowcroft, 1996). The goal of this chapter is to explore the particular implications that these partnerships have for the pursuit of sustainability. Sustainability is defined here as having both an economic and an environmental component. It refers to the simultaneous preservation of 1) the present course of economic expansion and capitalist accumulation and 2) the maintenance of the

T.J.N.M. de Bruijn and A. Tukker (eds.), Partnership and Leadership, 41–59.

ecological conditions necessary to support human life at a certain level of well being through future generations (O'Connor, 1994; Baker et al., 1997).

This chapter approaches this topic from a distinctively anthropological orientation. In particular, it directs its attention toward the multiple meaning systems that come together in these collaborative processes and the ways that these meanings are negotiated in practice. Its thesis is that certain ways of knowing and acting tend to be 'privileged' over, i.e. found to be more 'appropriate or efficacious' than (Wertsch, 1991:124), others in these settings and that this affects what these initiatives ultimately can and can not do with respect to fostering sustainability.

The discussion commences with a brief description of multistakeholder environmental partnerships and the research upon which this investigation is based. Next, the focus turns to a key finding from this research: the discovery of the presence of a dominant discourse in the partnership setting. This is followed by an exploration of the implications that this finding has for the pursuit of sustainability.

2. MULTISTAKEHOLDER ENVIRONMENTAL PARTNERSHIPS

Several main features distinguish the multistakeholder environmental partnerships that serve as the focus for this study. First, these partnerships assemble representatives from minimally the business, government, and non-governmental organisation (NGO) sectors for the purpose of addressing particular environmental concerns. To participate, organisations and individuals must qualify as 'stakeholders'. In other words, participants must be affected by or capable of affecting in some significant way the particular environmental planning, management, or policy decisions at issue. Second, these partnerships arise not out coercion but instead out of a mutual desire to address common environmental problems in a voluntary, proactive, and collaborative fashion. Third, the goals of partnerships are typically cooperatively defined. These goals may range from the production of policy recommendations to the development and implementation of contractual accords between the regulating and regulated communities. Finally, partnership processes proceed via consensus decision making, where responsibility for actions taken is shared among all of the participating actors.

3. RESEARCH DESIGN

This study draws upon ethnographic research performed between 1994 and 1998 with two environmental partnerships: a European Union (EU) level initiative aimed at enabling the practice of sustainable development in Europe, and a national-level project directed toward bettering natural resource management in the United States (U.S.). These cases derive from a larger 'collective case study' research project focusing on the practice and products of environmental partnerships (Stake, 1994:237).[1] In this chapter, the two case study partnerships are utilised 'heuristically' in the hopes of stimulating new understandings of similar types of processes (George, 1979:51).

3.1 Theoretical orientation

To date, most analyses of multistakeholder collaborative processes have tended to explain behavior in these settings in terms of the rational pursuit of self-interest. Such rational choice type approaches have been criticised, however, for decontextualising people's actions (Bourdieu, 1977) and for ignoring some of the socially and culturally based factors involved (Green and Shapiro, 1994; Schwarz and Thompson, 1990). This research draws instead on a practice theory perspective (Bourdieu, 1977) in order to draw attention to what I call the *sociohistoric* differences at play in these collaborative endeavours and to situate these differences within specific contexts of unequal and evolving relationships of power. These sociohistoric differences refer to the varying understandings of the environmental problems at issue that participants bring to partnership processes, the divergent ways by which these actors attach significance to these problems, and the differing types of actions they commonly pursue to resolve them. These conventional understandings are historically derived, having been developed in past institutional activity on and struggle over environmental issues. They are also expressed in the form of different types of *discourses,* defined as particular ensembles of ideas and categorisations through which meaning is given to physical and social realities (Hajer, 1995). These discourses privilege certain interpretive frameworks and are themselves more or less privileged by those participating in the partnership process. This sociohistoric-based approach thus encourages us to see collaborative processes as encounters between differentially valued meaning systems rather than merely as confrontations among competing interests.

[1]See Poncelet (forthcoming) for a more complete analysis of the collective case study.

3.2 Methods

My method for studying the case study partnerships was ethnographic - a method that emphasises the systematic description of social and cultural processes based on first hand observation and participation (Atkinson and Hammersley, 1994). Data were collected via participant-observation of all partnership meetings and collective activities held during the research period, the assemblage of textual materials produced by the partnerships and their participant organisations, and semi-structured interviews with a broad sample of partners. The interviewees were strategically selected so as to adequately represent the three major sectors, the variation in size of the participating organisations, and the range of level of participation in the partnerships. Approximately 25 interviews, each lasting around 90 minutes, were conducted for each partnership. The interviews were open-ended and designed so as to elicit naturally occurring speech on a range of topics, including personal and organisational background, multistakeholder collaborative processes in general, and the case study partnership in specific. The discourses observed were inferred via the discourse-analytic technique of focusing on key metaphors, terms, repetitions, generalisations, and evaluative statements (Price, 1987; Quinn and Holland, 1987).

3.3 Case study partnerships

Research with the EU case study, which I call the European Union Partnership for Environmental Cooperation (a pseudonym),[2] or EUPEC, was performed over a twelve-month period in 1994-1995. The EUPEC partnership was comprised of over 50 members representing multi-national corporations, national and European level environmental NGOs, sub-national level public authorities, trade unions, consulting groups, and think tanks. Established shortly after the 1992 United Nations Conference on the Environment and Development, it grew out the combined efforts of two leaders from the environmental and business communities in Europe to create an opportunity for their two sectors to meet. Its official mission was, first, to stimulate dialogue and cooperation between all sectors involved in or affected by the implementation of the EU's Fifth Environmental Action Programme and, second, to serve as a model for future partnership initiatives at the national, regional, and local levels. It also had the unofficial goal of utilising multistakeholder collaboration to produce policy recommendations

[2]As is typical of anthropological research, pseudonyms are used for the partnerships, organizations, and individuals discussed in this chapter. Protecting the identities of participants and their organizations serves to elicit greater freedom of expression in the meeting and interview settings.

on issues pertaining to sustainable development. To date, it has addressed environmental concerns in such pertinent areas as transport, tourism, agriculture, and water quality.

Structurally, the EUPEC partnership consisted of a general assembly, made up of all of the partnership's members, and an executive committee, comprised of equal numbers of participants from the NGO, business, and governmental sectors. The executive committee was responsible for coordinating partnership activities, including organising informational seminars and partnership workshops and establishing expert groups to address key environmental issues of common concern. These expert groups were comprised of interested EUPEC members as well as technical specialists brought in from the outside. Membership dues were the primary source of financial support for the partnership, although funding from the European Commission was also significant. Participation was open to any organisation, but official membership required approval by the EUPEC general assembly.

The U.S.-based case study, which I have entitled the Collaboration for the Improved Management of Natural Resources, or CIMNR, was created in the mid 1990s with the intent of exploring alternatives to the conventional system for regulating industrial pollution in the U.S. During my two years of research with the project in 1997-1998, its goal had shifted toward the development of improved approaches for managing natural systems. It was comprised by 40 or so participants representing large resource intensive and extraction-based industries, a range of national and sub-national level environmental advocacy organisations, and regulatory agencies and political officials at the federal and state levels.

The CIMNR initiative originated out of the growing concern on the part of a small group of business representatives that the existing system of environmental management in the U.S. was excessively expensive and inefficient. Convinced that an improved system could only be developed via collaboration among a wide range of stakeholders, they assembled a small group of interested industrialists, environmentalists, and public authorities and then sought out the support of a neutral coordinating body to help convene, administer, and facilitate a broader stakeholder process. Once a larger, more balanced partnership was established, it began meeting quarterly for three days at a time.

Financial support for the first year of the partnership was assumed by one of the original business representatives. After that, the initiative was funded primarily by donations from the major corporate participants and by grants solicited from several philanthropic foundations and the U.S. Environmental Protection Agency. Unlike the EUPEC project, participation in the CIMNR partnership was by invitation only.

4. THE PRIVILEGED DISCOURSE OF ECOLOGICAL MODERNISATION

At the heart of this analysis of the relationship between environmental partnerships and sustainability lies the finding that one discourse in particular was being accorded more legitimacy and status than any other in the case studies examined. This was the discourse of *ecological modernisation.* Three major premises form the basis of the ideology underlying this discourse (Blowers, 1998; Hajer, 1995). The first is the assumption that current environmental problems may largely be resolved via technological change and improvements. The second is the idea that long term economic development and environmental protection are interdependent and in need of integration. Ecological modernisation rejects any assumed fundamental opposition between economy and environment, promoting them instead in a positive sum relationship. The concept of 'sustainable development' and the dictum that 'pollution prevention pays' nicely express this logic. The third premise involves the presumption that the best way to achieve economic and environmental integration is via collective, concerted actions on the part of a broad range of societal actors. The discussion in the sections below focuses primarily on the last two of these premises.

4.1 Evidence of ecological modernisation

Numerous examples from the interviews, partnership publications, and meetings themselves illustrate the use of the discourse of ecological modernisation. For instance, the assumed need for better integration of economic and environmental goals is evidenced by the following comments of a EUPEC participant:

> "I think the benefits [of participating in EUPEC] lie in...better managing the economic-environmental ties - in integrating the economic dimension into environmental concerns. I believe that only by forcing the environment to leave its narrow preoccupations and integrating it into the economic machine will the environment reach fulfilment."

Even more profuse were statements alluding to the perceived positive sum relationship between economic and environmental forces. This is illustrated by the repeated promotion of the concept of 'sustainable development' as a major environmental goal as well as by the numerous comments made regarding the need for 'win-win' solutions. These ideas are evident in the following excerpt from a EUPEC publication on partnership and shared responsibility:

> "Partners will succeed together and the objectives of Sustainable Development will come to fruition when investments are directed to business opportunities that create jobs and measurably improve environmental quality. The triple win of environmental quality, business opportunities and job creation will be the challenge of the participants in the [EUPEC] programme and initiatives."

Ecological modernisation's promotion of collective approaches to environmental action is exhibited by frequent comments linking multisectorial dialogue and collaboration to the successful resolution of contemporary environmental problems. Some participants displayed this aspect of the discourse when speaking of multistakeholder partnerships as a generally superior way of taking action. Such was the case when a EUPEC participant stated:

> "Partnership is a way...is a tool for getting to a better end result faster. ...It reduces waste [in the form of] time and money and effort."

Other participants contrasted collaborative processes with what they considered to be less effective adversarial approaches. Along these lines, a CIMNR representative asserted:

> "There is a general agreement that the traditional command-and-control regulatory system, with the stakeholders participating in legal hearings or court-like procedures, is just not going to work in the future, and that collaboration, partnership, stakeholder types of common efforts [are] going to be a major element of a better way of doing things in the future."

Still others went even farther, making stakeholder collaboration a precondition for achieving sustainable development. The EUPEC coordinators demonstrated this view when they wrote in one of their publications:

> "[The EUPEC partnership] is not a group to lobby or defend positions. Its only bias is a firm belief that partnership actions will be more effective in the long run to ensure the conditions of Sustainable Development."

4.2 The privileged discourse

The use of the discourse of ecological modernisation, however, does not in itself substantiate its privileged status. To demonstrate this, we must situate its use within the actual practices of these two partnerships. In this section, the predominant position of ecological modernisation is revealed by its capacity to generate common ground in the fact of conflict. Two examples of contestation from the two case studies serve to illustrate.

Debate over economic reform in the EUPEC partnership

A first example comes from the EUPEC partnership and concerns a conflict that took place in an executive committee meeting. The committee was discussing two versions of a proposed business plan intended to guide the partnership. One version had been written by representatives from the business sector (led by Laurent), while the other had been drafted in response by two members of the NGO community (Stephan and Marcus). The conflict centred around two specific sections of the proposed business plan: one entitled 'The EUPEC Partnership's Unique Contribution,' and the other labelled 'The Endproduct'. The original business sector version as well as the NGO suggested replacement of these two sections are reproduced in Table 1 below.

Table 1. Comparison of proposed EUPEC Business Plans

The EUPEC Partnership's Unique Contribution:	
Business version:	NGO version:
[The EUPEC partnership] brings sectors of the European society (NGOs, business, legislators, unions, academia, etc.) together in a very practical setting and provides the know-how, conditions and *trust* to innovate together. The deadlock of perceived conflicting interests between sectors of society transforms into the challenge of *sharing responsibilities* and creating commitment for the effective implementation of *innovations towards sustainability* in a *socially attractive manner* (emphasis mine).	By successfully implementing innovative and *fundamental changes* (i.e., executing the projects) gradually a unique network of individuals and organisations will shape itself, with a mutual fascination for *breakthrough innovations* that lead to a "*reinvented economy*" (the way we consume, produce, work, manage, etc.). And with a feeling of being part of a fascinating, not formalised, alliance, which [sic] virtual bonds were created by shared intense learning experiences (emphasis mine).
The Endproduct:	
Business version:	NGO version:
[The EUPEC partnership]'s efforts will lead to actions supported by the [EUPEC] Membership with a *measurably positive environmental effect*. This can take the form of recommendations to the EU administration on regulatory matters, campaigns to educated consumers, jointly run experiments on business practice etc. (emphasis mine).	[The EUPEC partnership]'s efforts will lead to projects supported by the [EUPEC] Membership to build a *new European development model*, to help make *break-through innovations* in business practice, consumer behavior, NGOs, public authorities, trade unions, institutes approaches (emphasis mine).

In his presentation of both the original business plan and the proposed NGO revisions, Laurent stated that with regard to the EUPEC partnership's unique contribution:

> "There is a difference there between what I am proposing and what is written here [in the NGO version]. ...The document that [Stephan] proposes has a 'reinvented economy' and ways we consume, produce, etc. as an end result that to me sounds a bit like revolution and...and losing sight of the objective which is: we want to have a sustainable Europe, not how we want to get to a sustainable Europe by creating a revolution, but we want to get to a sustainable Europe. The difference here is that I'm trying to define the end result. ...And the same comments essentially apply to the proposed recommendation for the Endproduct. It concentrates on a 'new European development model' to help break up...whatever exists and replace it by something else. ...I'm much more in favour of defining the end result than in defining the way toward the end result."

At this point, the Dieter, a representative from the government sector, added:

> "The Endproduct: I think there could be said something more, in the line of what [Stephan] has put, but not necessarily in revolutionary terms. It could say something more about sustainability or what is meant by "measurably positive environmental effect"."

David, another leading representative from the business sector in the environmental arena, concluded the conversation with the following:

> "I can [identify] with some of the comments from [Stephan and Marcus]. I would like to see some more ambition of...a benefit to society and then a benefit to membership in the vision. And I think it's almost all there. It's just a matter of putting sustainable development and the Fifth Environment Action Programme rather up front. That is what should energise us to do something. And reveal in the mission [statement] the key process by which we do it, which is create partnership on priorities we decide together."

Stephan was present at this meeting yet elected not to respond to these arguments. This inaction is significant in itself as Stephan, an active leader in the European environmental movement, was frequently writing and speaking passionately on the need to transform Europe's economic structures in other social venues.

In interpreting this debate, I suggest that we view the discussion as an interaction between two competing discourses. Stephan and Marcus' arguments are grounded in ideas and assumptions subsumed under what Dryzek (1997) has termed 'green rationalism,' while the responses from Laurent, Dieter, and David, representatives in this case from the business and governmental sectors, are firmly rooted in the ideology of ecological modernisation.

According to Dryzek (1997), green rationalism is one of a number of discourses found within the environmental community. It shares with most of these the presumption that resource exploitation and environmental degradation have finite limits. Where green rationalism differs from other environmentalist discourses is in regard to its proposed solution to current environmental crises. Drawing on a critique of instrumental reason and a belief in the pre-eminence of human equality, green rationalism assumes that the resolution to the multi-faceted social and economic issues underlying today's environmental problems lies in radical political action and structural change. It calls for the establishment of more egalitarian political institutions and questions the suitability of the dominant capitalist system. Here is where the discourse of green rationalism contrasts the most with ecological modernisation. While ecological modernisation acknowledges that current environmental problems are themselves products of the prevailing capitalistic system of production and consumption, it does not conclude that this system must therefore be replaced. On the contrary, ecological modernisation views existing political, economic and social institutions to be the most appropriate structures for addressing issues of environmental protection (Hajer, 1995).

What I suggest took place in the interaction under examination is that when the two competing discourses confronted one another, the discourse of ecological modernisation prevailed due to its capacity to provide common ground for all of the contestants. When the business and governmental representatives drew upon ecological modernisation to respond to Stephan and Marcus' green rationalist calls for a 'reinvented economy' and a 'new European development model,' they shifted the debate away from its real area of disagreement, that is, over the contributing role of liberal capitalism to current environmental problems and toward a goal shared by everyone in the EUPEC partnership, that of a 'sustainable Europe.'

Debate over the role of stewardship in the CIMNR partnership

A second example illustrating the privileging of ecological modernisation comes from the CIMNR case study. The conflict here centred on a recurring debate over the utility of the concept of 'stewardship' as a means for improving natural resource management in the U.S. One particular manifestation of this conflict grew out of an argument between two individuals: Darcelle, an environmental manager from a large corporation in the energy sector; and Kirk, a representative of an international environmental organisation. Darcelle initiated the dynamic when she stated that instead of being rewarded for attempting to be a 'good steward,' current regulatory and economic incentive systems were such that her company was all too often penalised instead. This statement visibly disturbed Kirk, and he asked Darcelle what kind of compensation her firm wanted. Was it just public recognition? Darcelle responded that it was broader than that. She wanted the financial community to value her corporation accordingly. Kirk responded somewhat derisively that this was purely an economic motivation.

The meeting facilitator, sensing a brewing conflict, stepped in at this juncture and pointed out that Kirk and Darcelle have different understandings of stewardship. According to the facilitator, Kirk sees stewardship as an ethic, as something people do voluntarily because of its intrinsic value. For Kirk, the idea of receiving economic compensation for doing the right thing goes against this ethic. Darcelle, on the other hand, sees stewardship more along the lines of a practice. She has no problem reconciling profit-driven behavior and stewardship so long as the behavior accomplishes the natural resource conservation goals of stewardship.

Shortly after this interlude Darcelle picked up the discussion again, arguing that people need to be more creative in their attempts to promote stewardship. She described how her organisation, in collaboration with a prominent environmental group, developed an energy efficiency program that had the end effect of decreasing resource consumption while at the same time earning a profit. At this point, Ken, a representative from a federal regulatory agency, objected to Darcelle's position, stating that the partnership's goal is not to defend corporate bottom lines but to protect the environment. Another environmental NGO member, Cathy, spoke up here declaring that she too rejected Darcelle's definition of stewardship. Echoing Kirk's position, Cathy contended that stewardship ultimately originates out of enlightened self-interest and proceeds in the form of short term sacrifices for long term gain.

Eric, a representative of the environmental group that had worked with Darcelle's company on the energy efficiency project, came to Darcelle's defence at this point, cautioning against her opponents' line of reasoning. He stressed that the CIMNR partnership needs to break through this 'trade-off paradigm'. Instead of asking for environmental benefits in exchange for economic sacrifice, the group should be asking itself: How can we achieve better stewardship of resources *and* have it be profitable? Kirk and Cathy continued to resist this line of reasoning, however, replying that the energy efficiency project was not about stewardship but about perverse incentives and regulatory barriers. Kirk, in particular, was still irked by the idea that corporations will only do the right thing if they are compensated financially for doing so.

Darcelle re-entered the debate here. Picking up from Eric's comments, she replied that for her, being a good steward means that the impacts of her company on nature are not 'negative' but 'sustainable'. She also felt that society should value such actions by her company, and not only economically. At this point, Mark, an influential representative of the governmental sector, closed the discussion by stating that in the context of sustainability, stewardship includes not only resources but also human beings and livelihoods. He added that it is therefore hard to eliminate the economic component even if people want to do so.

The following day, representatives from both sides of the conflict gathered together in a working group with the goal of producing a common definition of stewardship. At the end of their meeting, they concluded that stewardship needs to be understood as being influenced not only by individual and group moral responsibilities but by economic and environmental policies as well. They recommended to the plenary that stewardship be understood as both an ethic and a practice, because only in this way may it be prioritised to achieve 'win-win' success factors.

In interpreting this interaction, I suggest that we may, as with the first example, view it as an instance of a radical green discourse being trumped by ecological modernisation. One side of the conflict drew upon the discourse of green rationalism and, in particular, the green rationalist critiques of the instrumental reason and anthropocentric arrogance underlying assumptions of unlimited economic growth (Dryzek, 1997) to object to the profit-oriented nearsightedness of the business sector. The other side of the conflict responded to this attack by criticising assumed economic-environmental trade-off paradigms and calling instead for actions that do not penalise business but instead encourage the development of solutions to natural resource problems that have both environmental and economic benefits. As in the EUPEC case study, this side promoted the concept of 'sustainability' as the true goal of environmental management. The actual privileging of ecological modernization over green rationalism is demonstrated by the final recommendations drawn by the bipartisan working group. This group closed the issue by redefining stewardship as both an ethic influenced by environmental values and morals *and* a practice driven by economic and political constraints. In this way, both the economy and the environment were integrated into a new 'win-win' conceptualization of the term.

5. IMPLICATIONS FOR SUSTAINABILITY

Given this argument for the privileged status of the discourse of ecological modernisation in the multistakeholder environmental partnerships examined above, we may now turn our attention to the topic question of this chapter: What does this imply for the pursuit of sustainability? This section explores some of the advantages and disadvantages that emerge from this finding.

The primary benefit that such partnerships pose for sustainability is associated with the fact that sustainability is itself an eminently ecologically modernistic concept. Consequently, partnerships promoting ecological modernisation are also, by definition, forwarding the sustainability agenda of simultaneous economic and ecological advancement.

Other benefits also derive from specific contributions that ecological modernisation presents for environmental problem solving in general. For instance, ecological modernisation provides a way of overcoming some of the entrenched conflicts and gridlock that have hampered environmental reform in the past. By advocating the pursuit of 'win-win' scenarios, partnerships are able to induce stakeholders from business, government, and the environmental community, and especially those who have been able to see only economic-ecological tradeoffs in the past, to find creative solutions to environmental problems in which everyone's interests are served. This, in turn, encourages the formation of a whole new set of environmental behaviors from these actors. This includes industrialists who may begin to look at environmental protection less as an economic constraint and more as a means of making money. It also includes environmentalists and regulators who may start to view corporations not only as the cause of pollution problems but also as part of the solution.

These shifts may then lead to yet another benefit of ecological modernisation: improved relations among stakeholders in the environmental arena. These relations are founded upon good will and the shared understandings afforded by a common discourse rather than upon the mistrust and miscommunication that has marked many intersectorial interactions in the past.

Despite such benefits, however, the promotion of ecological modernisation within environmental partnerships also poses certain problems for the pursuit of sustainability. Blowers (1998) points to one disadvantage when he notes that ecological modernisation applies better to certain environmental problems than to others. Its technological reforms, for instance, are better suited for addressing the pollution prevention problems facing Western democracies than the nature conservation dilemmas more common to developing countries. It is also less appropriate for environmental problems that involve long term, high consequence risks and around which substantial scientific certainty presently exists (e.g. global warming and nuclear waste disposal). Finally, ecological modernisation is largely limited to situations marked by relatively high levels of consensus and social cohesion and by a generally perceived compatibility between economic and environmental goals. This means that ecological modernisation approaches may be less applicable in contexts where divergent interests, entrenched conflicts, and socio-economic inequality prevail.

A second disadvantage concerns the complicit nature of ecological modernisation's relationship to the system of liberal capitalism. As noted above, while ecological modernisation acknowledges the contributions of capitalist modes of production and consumption to current environmental problems, it also relies exclusively upon this system to resolve them (Hajer, 1995). Consequently, as a discourse, ecological modernisation supports the prevailing power structures around which liberal market economies, such as those of Western Europe and North America, are based. The problem here is that the dominance of this discourse has the effect of impeding critical examination of and debate over possible impediments that underlie such prevailing structures. Criticisms based upon the arguments that the market, alone, is incapable of protecting the environment or that liberal capitalism is itself not ecologically sustainable will carry little weight in these settings.

A third disadvantage to be considered stems from the very presence of discourse hierarchies in environmental partnerships. In particular, problems arise from the asymmetrical effects associated with adopting versus rejecting the dominant discourse. By assuming or promoting the dominant discourse, participants have the capacity to gain legitimacy, credibility, and status within the group. In the case of ecological modernisation, it is thus possible for actors who have traditionally been weakly positioned in the environmental or economic domains to improve their standings in these domains simply by espousing this discourse. Consider the example of a firm from a heavily polluting industry. Although this firm may have been an environmental laggard to date, by championing the favored discourse of ecological modernisation, it may readily redefine itself as an environmental leader both in the partnership and, by extension, in society. The problem with such instances, however, is that discursive practices alone do not necessarily translate into improved environmental or economic behaviors. The adoption of the discourse of ecological modernisation may thus serve to improve a company's green image, thereby enhancing its economic sustainability, while doing little or nothing on behalf of ecological sustainability.[3]

[3]A parallel argument could be made for an environmental group whose actions have had the effect of stymieing economic development in the past. Adoption of the discourse of ecological modernization would help such an organization to redefine itself as economically progressive while not guaranteeing any contribution to economic sustainability.

Paralleling the potential gains to be made by adopting a dominant discourse are drawbacks associated with choosing not to do so. This is because alternative discourses, and the participants who espouse them, will tend to be accorded less significance and granted less consideration by the majority others. The problem here is that if participants in multistakeholder processes stop listening to or offering competing perspectives, these perspectives risk becoming in effect suppressed.

This suppression of alternative voices occurs in a variety of ways. First, it takes place when participants advocating minority views in a collaborative process elect to stop talking. This is what happened in the EUPEC business plan conflict involving the attempts of an influential environmental NGO representative to introduce elements of the more radical green rationalism discourse into the debate. In this case, pressure placed upon him by other participants to relinquish his calls for a 'reinvented economy' in favour of the more broadly accepted goal of 'sustainable development' subsumed under the dominant discourse of ecological modernisation caused him to abandon his position. The result was that some of his more 'out-of-the-box' ideas faded from the discussion.

Secondly, alternative voices are suppressed when the stakeholders espousing them opt to terminate their participation in partnership processes. Over the course of its operation, the CIMNR initiative suffered the departure of stakeholders at both ends of the ideological spectrum. On the one hand, this involved several representatives from the environmental justice movement. These participants, whose primary interest lay in defending the plight of the many socio-economically disenfranchised segments of society currently being forced to accept a disproportionately large share of the burden of environmental degradation, chose to stop coming in part because they felt that their attempts to inject an environmental justice perspective into the proceedings were being disregarded and ignored. On the other hand, the partnership also lost an industrial participant espousing the ideology of free market environmentalism, a perspective that advocates the privatisation of property and the relaxation of regulatory controls so as to enable the market to automatically inhibit the negative effects of environmental pollution and resource exploitation. This individual again attributed his departure in part to the unwillingness of the other participants to engage his ideas. He expressed specific frustration with the reluctance of CIMNR members to critically reflect on the presumption of resource scarcity embedded in the discourses of both ecological modernisation and green rationalism - an idea directly challenged by the free market environmentalist assumption of resource abundance. So he left, taking his rather innovative ideas with him. His departure, along with that of the environmental justice representatives, left the CIMNR partnership even more entrenched in the mainstream middle.

A third way by which discourses may be suppressed in collaborative processes occurs when legitimate stakeholders promoting alternative views choose not to participate in the first place. In the EUPEC partnership, this happened when a prominent environmental NGO with a reputation for being radical rejected an invitation to join. This NGO announced in a decidedly non-ecologically modernistic manner that it objected to both the concept of sustainable development and the idea of multistakeholder collaboration. As a result, the EUPEC partnership lost an important opportunity to expose itself to a broader range of environmentalist views.

Out of this third overall disadvantage that environmental partnerships privileging the discourse of ecological modernisation pose for the pursuit of sustainability comes a fourth and final limitation to be considered. At issue here is the matter of economic-ecological equity. In particular, the manner in which the discourse of ecological modernisation was able to trump its radical green counterparts in the EUPEC and CIMNR case examples presented above raises the question of whether environmental interests are benefiting to the same extent as business interests in these settings. It also forces us to recognize that not all stakeholders will necessarily share equally in the alleged 'wins'. At greatest risk of losing out are those stakeholders who do not subscribe to the basic premises of ecological modernisation. These include representatives from the more radical wing of the environmental movement that operate by confrontational means rather than by cooperation, that question the suitability of socio-economic systems grounded in liberal capitalism, and that insist on addressing the natural environment from a moralistic rather than an economistic perspective. Even for those stakeholders who advocate the ideals of ecological modernisation, there is no assurance that the ecological gains achieved in such partnerships will be as substantial as the economic ones.

6. CONCLUSION: ON THE SEARCH FOR THE 'WIN-WIN'

In the abstract, one might argue that multistakeholder collaboration, by definition, leads to 'win-win' situations. Any time societal actors choose to work with rather than against one another to resolve common problems, the chances are improved that they will arrive at solutions more advantageous to all (Driessen, 1998). In the context of the ideology of ecological modernisation, the collaborative production of 'win-win' solutions takes on a more specific meaning, involving the co-achievement of economic prosperity and environmental protection.

This chapter presents evidence demonstrating the privileged status of this ideology in certain environmental partnership settings. It also explores some of the advantages and disadvantages associated with this finding. From this discussion, we may conclude that multistakeholder partnerships privileging the discourse of ecological modernisation are likely to have contradictory implications for sustainability. On one hand, these partnerships promote an ideology and strategic approach that values and fosters economic-ecological 'win-win' solutions to contemporary environmental problems. This lies at the very heart of the sustainability agenda. On the other hand, by limiting the topics open for discussion and by suppressing alternative perspectives, these partnerships also serve to restrict what these 'win-wins' may accomplish. This suggests that while such partnerships constitute a step toward sustainability, they will likely not be enough to assure the achievement of this goal at a societal level. These partnerships may even render a disservice if, in prioritising 'win-win' fixes, they cause us to forget that not all 'wins' are the same and that there will often be losers.

ACKNOWLEDGEMENTS

The research on which this chapter is based was funded by The Belgian-American Educational Foundation, University of North Carolina Graduate School and Sigma Xi - Grant-in-Aid of Research

REFERENCES

Anderson, Terry L. (1997) Prologue, in *Breaking the Environmental Policy Gridlock*. T. Anderson, ed. Stanford, CA: Hoover Institution Press.

Atkinson, Paul and Martyn Hammersley (1994) Ethnography and Participant Observation, in *Handbook of Qualitative Research*. N. Denzin and Y. Lincoln, eds. Thousand Oaks, CA: Sage Publications.

Baker, Susan, Maria Kousis, Dick Richardson and Stephen Young (1997) Introduction: The theory and practice of sustainable development in EU perspective, in *The Politics of Sustainable Development: Theory, Policy and Practice within the European Union*. S. Baker et al., eds. London and New York: Routledge.

Blowers, Andrew (1998) Power, participation and partnership: The limits of co-operative environmental management, in *Co-operative Environmental Governance: Public-Private Agreements as a Policy Strategy*. P. Glasbergen, ed. Dordrecht: Kluwer Academic Publishers.

Bourdieu, Pierre (1977) *Outline of a theory of practice*. Cambridge: Cambridge University Press.

Buttel, Frederick H. (1992) Environmentalization: Origins, Processes, and Implications for Rural Social Change. *Rural Sociology* 57(1): 1-27.
Collier, Ute (1998) The environmental dimensions of deregulation: An introduction, in *Deregulation in the European Union: Environmental Perspectives*. U. Collier, ed. New York and London: Routledge.
Driessen, Peter (1998) Concluding remarks: The scope of co-operative management, in *Co-operative Environmental Governance: Public-Private Agreements as a Policy Strategy*. P. Glasbergen, ed. Dordrecht: Kluwer Academic Publishers.
Dryzek, John S. (1997) *The politics of the earth: environmental discourses*. Oxford: Oxford University Press.
Fiorino, Daniel J. (1996) Environmental Policy and the Participation Gap, in *Democracy and the Environment: Problems and Prospects*. W. Lafferty and J. Meadowcroft, eds. Brookfield, VT: Edward Elgar Publishing Company.
George, Alexander L. (1979) Case Studies and Theory Development: The Method of Structured, Focused Comparison, in *Diplomacy: New Approaches to History, Theory, and Policy*. P. Lauren, ed. New York, NY: The Free Press.
Gill, Stephen (1995) Globalisation, market civilisation, and disciplinary neoliberalism. *Millennium: Journal of International Studies* 24(3): 399-423.
Golub, Jonathan (1998) New instruments for environmental policy in the EU: Introduction and overview, in *New Instruments for Environmental Policy in the EU*. J. Golub, ed. London and New York: Routledge.
Gray, Barbara (1989) *Collaborating: Finding Common Ground for Multiparty Problems*. San Francisco: Jossey-Bass.
Green, Donald P. and Ian Shapiro (1994) *Pathologies of Rational Choice Theory: A Critique of Applications in Political Science*. New Haven and London: Yale University Press.
Hajer, Maarten (1995) *The politics of environmental discourse: Ecological modernization and the policy process*. Oxford: Clarendon Press.
Lafferty, William M. and James Meadowcroft (1996) Democracy and the Environment: Prospects for Greater Congruence, in *Democracy and the Environment: Problems and Prospects*. W. Lafferty and J. Meadowcroft, eds. Brookfield, VT: Edward Elgar Publishing Company.
Long, Frederick J. and Matthew B. Arnold (1995) *The Power of Environmental Partnerships*. Fort Worth, TX: Harcourt Brace and Company.
O'Connor, James (1994) Is sustainable capitalism possible? in *Is Capitalism Sustainable? Political Economy and the Politics of Ecology*. M. O'Connor, ed. New York: The Guilford Press.
Poncelet, Eric C. (forthcoming) *Partnering for the Environment: Multistakeholder Collaboration in a Changing World*. Lanham, MD: Rowman and Littlefield Publishers, Inc.
Price, Laurie (1987) Ecuadorian illness stories: Cultural knowledge in natural discourse, in *Cultural Models in Language and Thought*. D. Holland and N. Quinn, eds. Cambridge: Cambridge University Press.
Quinn, Naomi and Dorothy Holland (1987) Culture and Cognition, in *Cultural Models in Language and Thought*. D. Holland and N. Quinn, eds. Cambridge: Cambridge University Press.
Schwarz, Michiel and Michael Thompson (1990) *Divided we stand: Redefining politics, technology and social choice*. Philadelphia, PA: University of Pennsylvania Press.
Stake, Robert E. (1994) Case Studies, in *Handbook of Qualitative Research*. N. Denzin and Y. Lincoln, eds. Thousand Oaks, CA: Sage Publications.

Wertsch, James V. (1991) *Voices of the Mind: A Sociocultural Approach to Mediated Action.* Cambridge MA: Harvard University Press.
Wondolleck, Julia M. and Steven L. Yaffee (2000) *Making Collaboration Work: Lessons from Innovation in Natural Resources Management.* Washington, D.C.: Island Press.

Chapter 4

RE-INTERPRETING LOCAL-GLOBAL PARTNERSHIPS

SUSSE GEORG

ALAN IRWIN

1. INTRODUCTION

Within international discussions of sustainable development, increasing attention is being paid to local initiatives designed to provide what is often presented as a 'bottom-up' approach to environmental action. This broad approach follows the general rhetoric of sustainability as expressed in the Brundtland Report emphasising the need for widespread participation and empowerment. Meanwhile, and somewhat counter to this approach, the process of institutional response and development has largely been dominated by national and inter-national bodies. Accordingly, many efforts towards local development have been framed within a wider policy framework set by centralised institutional authorities and, as in the case of Agenda 21, subsequently delegated to local authorities (Young, 1997; O'Riordan and Voisey, 1998).

Seen in this light, the bottom-up agenda which emphasises local participation can be in direct competition (and confrontation) with more traditional 'top-down' governmental strategies and environmental initiatives undertaken by 'big' business. However, within the business and environment literature a great deal of attention is being devoted to conceptualising and describing business interactions with stakeholders other than government (Freeman, 1984; Long and Arnold, 1994; Donaldson and Preston, 1995;

T.J.N.M. de Bruijn and A. Tukker (eds.), Partnership and Leadership, 61–76.

Hartman and Stafford, 1998; Harrison and Freeman, 1999). Often coined as a matter of engaging, interacting, managing and partnering, stakeholder dialogue is, depending on one's perspective on business, either considered to be 'good' because of its collaborative nature or as 'greenwash' since it represents nothing more than the co-option of environmental concerns.

This chapter reviews these conventional arguments carried through spatial metaphors such as 'bottom-up', 'top-down', 'local and global', and distinctions between 'grassroots' and 'industry' to suggest that such dichotomies need to be re-considered in both academic and policy-related terms. Implicit in these dichotomies lie two received images of how changes take place. It is either seen as some *planned* endeavour – as in the 'top-down' initiatives of government or industry – or as something *emergent* that grows from the 'bottom up' as in the case of grassroot environmental groups/organisations. This paper seeks through two cases to illustrate how these images of change are contradicted by practice, thus suggesting the need for more nuanced analyses of environmental change and of environmental partnership.

2. LOCAL VERSUS GLOBAL

The current level of enthusiasm for bottom up approaches appears to be premised on two inter-related lines of argumentation: a critique of 'top-down' policy approaches and institutional strategies, and, in direct contrast, the glorification of local initiatives. Criticisms of centralised, top-down, approaches have centred on a number of perceived weaknesses and problems (Irwin et al., 1994; Irwin, 1995; Jacobs, 1997; Marvin and Guy, 1997; Macnaghten and Urry, 1998). These can be extended and outlined as follows:

- the disempowering nature of, for example, many government initiatives so that local communities simply become the chosen sites for environmental change rather than being treated as independent agents with their own priorities and concerns;
- the characteristic insensitivity to local differences – policies from the centre typically emphasise similarities and generalised assumptions across locations;
- the narrow targeting of environmental (and efficiency) improvements in business without bringing about transformational or systematic changes in wider commercial operations;
- the separation of environmental factors from other social, political and economic concerns – so that policy agenda for different issues are set

separately rather than being brought together (or prioritised) within specific settings;
- the *constraining effects* of top-down policies so that alternative visions of environmental development are implicitly downgraded in favour of more conventional and incremental strategies;
- the creation of an *artificial distinction* between 'environmental' and other issues (rather than, for example, developing a wider vision of 'quality of life' as defined by particular communities).

Taken together, these criticisms and comments emphasise the importance of local context for environmental and social development. Rather than simply constituting sites for the implementation of environmental policies decided elsewhere, the argument is that such contexts represent significant settings for the development (as well as application) of environmental policies and strategies. In making these arguments, a case is also being presented for a more democratic approach to environmental strategy: local initiatives offer greater potential for citizen empowerment and participation precisely because they provide contextually appropriate modes of social articulation and ecological change. In so doing, local initiatives combine both a 'democratic' and an 'ecological' rationale.

In contrast to centralised approaches, 'bottom-up' initiatives such as business-environmentalist partnerships as well as community or local governmental efforts to improve specific environments are presented as offering distinct advantages (Jamison et al., 1997; Mander and Goldsmith, 1996; Szerszynski, 1996; Lerner, 1997):
- they relate more closely to citizen concerns and hence offer a *more accessible and equitable approach*;
- they *channel and enhance* the diversity and imagination demonstrated by citizen-led campaigns and initiatives, and allow for the development of locally adopted methods of resource use and knowledge;
- they represent an often-latent *environmental resource* in the form of local people, local concerns and local expertises;
- rather than suggesting an artificial separation of topics and possibilities for action, they allow environmental concerns to be *placed alongside* a range of local issues.

Taking these points together, local initiatives can be presented as an important focus for sustainable environmental activity. From this perspective, such initiatives access what can be termed a 'latent environmental need' – empowering citizens to influence industry and perhaps even develop their own environmental solutions that have not previously been on the market, 'environpreneurialism' (Hartman and

Stafford, 1998). Furthermore, such initiatives seem to build upon public expertises and offer a longer-term route to social and technological development. Within this framework also, centralised approaches can appear unwieldy and remote – as ponderous Goliaths out of touch with particular groups of citizens and their needs. Alongside the advocacy of local initiatives, therefore, can also be found a portrayal of centralised institutions as struggling with the environmental agenda for very deep-seated social and technical reasons.

As a counter-force to central initiatives, local approaches would appear then to have much to offer. At a more theoretical level, such approaches link to Ulrich Beck's analysis of the condition of late-modernity. Environmental issues raise wider concerns over social direction and over the tools and forms of expertise (including scientific knowledge) which are available to us. Accordingly, modernistic institutions (such as governments and industries) struggle to deal with the emerging conditions of the 'risk society' since they rely upon the very kinds of institution and policy instrument that are becoming outmoded and anachronistic in the contemporary social world (Beck, 1992). Meanwhile, a new form of 'sub-politics' is better-placed to construct 'pragmatic alliances' in the face of social and environmental uncertainty. Local initiatives – including partnerships between business and environmental groups – seem ideally positioned to bring together a diversity of groups with objectives which, at least temporarily, converge. Indeed, this is heralded as a new mode of politics grounded within the realm of civil society rather than in market and state institutions (Eder, 1996). Moreover, these local initiatives and partnerships can be seen as representing an organisational diversity that is important for the ability of societies to respond to pressing societal problems (Hannan and Freeman, 1989).

Despite the critique of centralised approaches and the rhetoric stressing the advantages of local initiatives, there is, however, still a general scepticism – especially within policy circles – as to the wider impact and viability of such initiatives. Implicit in their dismissal is that they are too small-scale and too limited in vision to entail significant environmental change. Moreover, local initiatives are often considered to be 'merely' an expression of utopian ideals, an expression of 'green romanticism' (Dryzek, 1997). Another common critique dismisses partnerships or 'green alliances' on the grounds that they primarily serve the interests of business and not that of the environmentalists.

Rather than continue this potentially fruitless debate over the pros and cons of 'bottom-up' versus 'top-down' strategies, in this paper we wish to take a more analytical look at relations between the 'local' and the 'global' within environmental initiatives. Going further, and rather than simply reifying either of these categories, we wish to consider their mutual construction within environmental discussion. Certainly, it seems impossible to define the 'local' without at least an implicit contrast with the 'global – and yet, as we will argue, this simple dichotomy blocks discussion of the cross-cutting and mutually-dependent relations involved. In particular, the reification of both the local and the global crucially fails to consider the basis on which factors are accorded either global or local status. As Yearley has put this point more generally:

> "'global environmental problems' are *global problems* because people and organisations have worked at making 'the environment' a world-wide phenomenon." (Yearley, 1996:143-4).

Moreover, these distinctions offer little in the way of understanding, for example, how and why certain ideas as to how to solve environmental problems emerge or 'travel' from one setting to another. In analytical terms it becomes important to view them as flexible constructions within particular contexts and settings. On that basis, we argue that terms such as 'local', 'global', 'bottom-up' and 'top-down' need to be seen as actively-generated categories rather than unproblematic descriptions of pre-existing entities.

In what follows we will delve into this issue of what constitutes the local by presenting examples of environmental initiative. We do so by telling two stories both of which can be said to be 'plain histories' (Law, 2000); moving chronologically and linking specific events and dates. They are, however, also policy narratives, highlighting things that – at least from our perspective – have policy implications. Both of these examples are taken from a previous European research project into the 'social management of environmental change' (see Irwin et al., 1994, 1995). We will start with an example of the development of an ecological village that fits well with the predominant conceptualisation of localism as having to do with spatial proximity and community development (Andersen, 1995). However, we will also see that this 'local' case depends crucially upon more 'global' agents – including the municipal housing society and a large, trans-national construction company.

We will then continue with a case that many would hardly consider as a local initiative, but as a typical example of a business-environmentalist partnership. This second example concerns the development of a now-commonplace household fixture - a CFC-free refrigerator (Verheul, 1995). Whilst international developments and large-scale industries play an important part in this story, we will also consider the importance of local and personal connections – and, notably, between German Greenpeace and two scientists.

There are particular themes associated with each of these examples that link the issue of localism to a larger context. What counts as local varies in the two accounts, yet in both instances, the stories entail processes and developments that transcend the localities in which they take place and lift the stories away from the physicality of 'merely' managing environmental resources locally. Taken together, these two examples allow us to discuss different constructions of 'the local' and of environmental partnership, and thus move beyond the simplistic portrayal of either local initiatives or partnerships as a panacea for all environmental ills.

3. A VILLAGE THAT IS A LITTLE DIFFERENT

Baalarna is a small Swedish village in the vast forests north of Stockholm, beautifully situated overlooking a lake. Twenty people, from five families, live in this idyllic spot. Judging from appearances, this village resembles many other small Swedish villages – consisting of a handful of red-painted wooden houses with white trimmings. There is, however, at least one thing that distinguishes Baalarna from other similar sized villages - it is an ecological village. It is also the first ecological village constructed by the local municipal housing society, 'Stiftelsen Nordanstigs Bostäder' (SNB).

The houses – despite their traditional appearance – are designed according to ecological principles, i.e. the materials used and the technical installations are all designed to minimise resource use. The windows, for instance, are made with three-layered glass as opposed to the normal (for Sweden) two. They have also been made larger on the south-facing facades and smaller on the north-facing, so as to maximise the use of passive heating whilst also minimising heat loss. The insulation is made of recycled paper that allows the houses to 'breathe' and heat is supplied by a wood-fired furnace. This is supplemented by solar heating, particularly in the summer. The sanitation system constitutes yet another unique feature. Installed in each house is a so-called 'separation toilet' which, as its name suggests, separates bodily wastes from one another. This has several advantages: water use is considerably lower than in a traditional flush toilet. The urine can be used as fertiliser. It is easier to treat the household's remaining 'grey' water by simply draining it into the surrounding fields. Moreover, this sanitation technology requires neither septic tanks nor sewage systems, both of which are difficult and costly to establish on bedrock and in a sparsely-populated area.

The village can, in some respects, be considered as a 'closed system' – and, therefore, as a clear example of the 'local' being separated from the 'global'. The villagers are self-sufficient with respect to water, which they get from a near-by lake, and they are partly self-sufficient when it comes to energy. They own the forest that surrounds them, and use the wood for heating, and they are able to manage most of their wastes. However, village life is less autonomous in other respects. The villagers do not, for instance, produce their own electricity, nor do they have any intention of becoming completely self-sufficient with regard to food. The village also houses a software company – representing one of the most 'global' of modern technologies. Viewed in this way, we can begin to see how a simple categorisation of this case as either 'local' or 'global' becomes difficult to maintain. Rather than representing either the local or the global, this case offers an imaginative and dynamic combination of these elements. This point can be reinforced by briefly considering the history of this social experiment.

The story of this eco-village dates back to the mid-1980s, when the five families began discussing the possibility of building their own homes. Prompted by the Chernobyl accident in 1986 and carried by a strong anti-nuclear sentiment, they wanted to build houses that could be as self-sufficient as possible with respect to energy. Moreover, they wanted 'healthy' houses, ones that could 'breathe' allowing the noxious and toxic vapours of the building materials to escape. Their plans were not carried by explicit ecological considerations. Indeed, they had no intention of establishing an ecological village – this 'just happened' in the wake of events that followed their collective decision in 1987 to go ahead with the project.

This soon proved to be no easy endeavour as it not only entailed locating the 'right' place and engaging an architect to help design the houses, but also finding the necessary financial resources. The latter proved to be particularly difficult. After almost three years of preparation, and just as the families were about to give up (in 1990), SNB – having heard of their problems – offered to organise and finance the project. They also suggested that the project be turned into an ecological village by integrating environmental considerations into the construction process. The village was built by Skanska, Sweden's largest construction company, after they won the call for tender in 1991.

This was, however, not a straightforward task – neither the villagers, the Director of SNB, nor the construction company had much prior knowledge of ecological house construction. Nevertheless, by drawing extensively on a network of professionals and grass-root organisations engaged in promoting ecological villages and technologies, the villagers, SNB and the construction company were able to tap into a much larger body of knowledge. As a Skanska representative put it: 'We're the ones learning'.

Building Baalarna also marked a change in how both Skanska and SNB work. It was the first time they had adopted the strategy of actively engaging with the 'users' in the development processes. Normally, residents would just move in once the houses are finished. However, in a situation like this, where the technologies either did not exist or were not easily accessible, co-operation was crucial for coming up with 'new' solutions. Heralded as a success, the village now is a show-case that can inspire others in their quest to develop sustainable technologies.

Rather than simply categorising this case as either 'local' or 'global' in character, we would argue that its interest and vitality reside in the creative *combination* of these elements through a lively and creative partnership of different actors and organisations. Whilst the villagers and their direct involvement in the design of their own homes serve to emphasise the 'local' dimension, the part played by the housing society and construction company suggests more 'global' elements as being at work. Equally, the kinds of knowledge and expertise being developed and applied in this case seem inextricably both local and global in character. Whilst the presentation of the Baarlana story as determinedly 'bottom up' in character captures some of its direction and motivation, such a characterisation significantly misses the wider links upon which this 'local' case depends: including connection to wider systems of knowledge, energy production and information technology. At the same time, our understanding of the character of environmental partnership is significantly broadened.

4. THE MAKINGS OF THE GREENFREEZE REFRIGERATOR

CFC-free refrigerators have become widely popular over the last few years. In buying these fridges consumers make – within the confines of their household – a local contribution to solving a global environmental problem. These refrigerators are less damaging than models using the ozone-depleting CFCs, and are now being produced by internationally-known appliance manufacturers, including Bosch-Siemens, AEG, and Electrolux. This has, however, not always been the case, and this story of how these fridges made it to the market provides us with another lens through which we can view the complex makings of environmental change.

In late 1993-early 1994 various major appliance producers announced that the coolants in most – if not all – of their coming refrigerators and freezers would be hydrocarbons as opposed to the fluorocarbons, HFCs (Haaland, 1994:6-11). This shift in production strategy can be attributed to a number of events that, in our account, take their start in 1988 at a medical research institution in Germany, The Dortmund Institute of Hygiene. Two scientists at the institute needed a new cooling system in their laboratory, but did not have the necessary funding. Instead, they constructed their own system; one that, much to the surprise of their colleagues, did not rely on

HCFCs or HFCs as coolants but on a combination of light hydrocarbons, including propane and butane which later became known as the 'Dortmund mixture'. Despite widespread scepticism within the scientific community, the scientists won an environmental prize in Germany for their endeavours, thus, attracting Greenpeace Germany's attention.

Neither government nor industry showed much – if any – interest in the scientists' initiative. There appears to be at least three reasons for this. First, early results suggested that refrigerators using hydrocarbon coolants consumed a great deal more energy than conventional ones. Second, the coolant's flammability was considered particularly problematic by many household appliance manufacturers. The third, and presumably most pervasive factor, was, however, the course of action taken by the chemical industry. By the late 1980s, and certainly with the signing of the Montreal Protocol (1987), the race was on to find substitutes for the environmentally-problematic CFCs, and the chemical industry had chosen to develop HCFCs and HFCs as alternative coolants. Moreover, they had also developed recycling schemes which 'locked' their customers – including the automobile and refrigeration/appliance industries – 'in' to these specific CFC alternatives (Gabel, 1995).

These developments were not, however, to Greenpeace Germany's liking. One of its main concerns was that the final phase-out of CFCs would be achieved by 'dropping in' HFCs and HCFCs. In early 1992, Greenpeace Germany engaged one of the scientists from the Dortmund Institute to continue testing the 'Dortmund mixture', and a few months later Greenpeace succeeded in convincing the former East German producer of household appliances Deutsche Kühl- und Kraftmaschinen (DKK Scharfenstein), known for its long-standing experience in producing CFC-free cooling systems, that they should join the project. Suffering from an enormous decrease in sales and a huge deficit following German reunification, the economic prospects – and indeed the survival prospects – for DKK Scharfenstein appeared bleak. According to the Treuhand, the state holding agency responsible for co-ordinating the sale of former East German companies, DKK's days were numbered, particularly as an attempt to sell the company to Bosch-Siemens-Hausgeräte GmbH had failed earlier in 1992. So, co-operation with Greenpeace was an offer they could hardly refuse.

This proved to be a fruitful partnership – DKK produced a prototype which Greenpeace then marketed broadly to consumers in an extensive publicity campaign. By the end of 1992, they had 65,000 advanced orders for the 'new' eco-fridge, 'Greenfreeze'. Moreover, Greenpeace marketed the fridge politically, soliciting the support for 'their' product from among others the German Minister of the Environment. Despite plans to liquidate

DKK, the Treuhand decided to let DKK continue. It invested several million DM in DKK before managing to sell it to The London East German Investment Trust, a consortium under the Berliner Bank, in August 1992. The company was re-named 'Foron', and began to set Greenfreeze into mass production in 1993.

By the following year, Foron had managed to gain a very small share of the German market. The refrigeration industry, consisting of six major companies – Bosch-Siemens, Liebherr, AEG, Electrolux, Bauknecht and Miele – was, however, a powerful player with which to contend. In an attempt to exclude Foron from the market they issued a communiqué to German retailers stating that the Greenfreeze refrigerator used more energy than conventional ones, that the hydrocarbon technology had not been sufficiently tested and calling Greenpeace irresponsible for promoting this technology (Haaland, 1994). However, with the introduction of Greenfreeze to the market, the refrigeration industry changed strategy and within a few months began to employ hydrocarbons and other substances as coolants in its refrigerators – something it had previously claimed it would take years to accomplish. According to the head of research and development in Bosch-Siemens, it was the public interest generated by the Greenfreeze campaign that proved a decisive factor in this change of industry strategy. In that way, the public interest and support generated by the Greenpeace campaign was "a very important 'eye-opener' for industry" (Verheul, 1995).

Compared to the case of Baarlana, this appears to be a much less 'local', and therefore more 'global', case of large-scale innovation and major companies. However, such a simple categorisation would again miss the point. On the one hand, one can identify 'local' elements within this ostensibly-global case: the local environmental development within a specific region during the German re-unification process, the particular link between Greenfreeze and these scientists, the reliance upon a small number of key actors. On the other, we find once again that it is the combination of local and global elements which makes this case so compelling: international protocols and a research institute, two scientists and the international refrigeration industry, a particular mixture of propane and butane, households and big business. In the face of such hybrid entities (Latour, 1993), the two-way classification into either the local or the global seems to ignore the most interesting and important dimensions of this case. Indeed, labelling of this kind misses the complexity of the processes involved in making environmental improvements.

5. RE-INTERPRETING ENVIRONMENTAL INITIATIVES

As should already be apparent, the point of this paper has not been to discuss these cases in any depth – nor to present them as in any sense 'typical' or 'representative'. Indeed, these stories could be told in any number of ways depending on the narrator's perspective (Czarniawska, 1997). What we wish to reflect upon is the wider implications for our understanding of 'local' environmental solutions, partnerships and other 'bottom-up' (or 'decentralised') approaches. Pulling together the previous discussion, a number of particular implications emerge.

Firstly, we can see that whilst each case depends to some degree upon local networks and contacts (the families in the first case, the particular scientists in the second), they simultaneously rely upon (and contribute to) global relations of knowledge and expertise. Moreover, such networks combine the local and the global rather than fitting neatly into either category. This account of local-global partnerships accordingly seems to offer a more compelling and convincing presentation of contemporary environmental change than the employment of more rigid and, as we have suggested, crudely-dichotomised terminology.

Secondly, and as also noted in both cases, it seems that the dynamism of such 'social experiments' stems (at least partly) from these innovative and imaginative combinations (separating toilets and a software company, two scientists and international companies). From a technological perspective, the narrow advocacy of localism seems to miss these new possibilities. The success of these initiatives is hinged on the heterogeneous engineering of a variety of different actors, engaged and enrolling each other in combining existing knowledges to produce new artefacts (e.g. toilets and insulation in Baalarna) and/or to develop new uses for existing artefacts (e.g. refrigerator coolants).

In cases such as these, it is almost as if the interesting action occurs *neither* at the local nor the global level but precisely at their point of *interaction and synthesis*. This directs attention away from rigid (perhaps 'purist') notions of 'local environmental change' and towards a greater awareness of the actor networks involved in mobilising and stabilising particular 'solutions' (Callon, 1986; Latour, 1993; Law, 1992). Whilst the call for localism can certainly represent a useful stimulus to action, it is also possible that it can divert attention away from analysing these possibilities.

Thirdly, a simple spatial notion of the local is inappropriate to these cases. Instead, it seems that the local can be at work in apparently-global industrial systems and the global within a particular locality. Viewed in this way, the language of the 'local' (and the 'global') represents a relationship and a perspective rather than a fixed entity. In that way also, talk of the 'local and global' has performative properties. For example, Greenpeace enacted a local solution at a global level. Players at this level continued, then, to mobilise the solution on the consumer market for very 'local' actors to buy. Equally, and rather than viewing particular developments as intrinsically 'local' or 'global' in character, there is a need for further exploration of localising and globalising forms of environmental discourse.

The terminology of 'bottom-up and top-down' encounters similar problems: the villagers' 'bottom-up' strategy can appear 'top-down' to those working for the housing society – and vice versa. Greenpeace, often criticised for being a top-down organisation, can engage in bottom up activities when it comes to engaging with the refrigeration industry. In all these cases, matters of definition appear more flexible and open to varying construction than the conventional form of discussion would allow.

Fourthly, and to hint at a parallel line of argument, it can also be observed that these cases do not separate social from ecological elements. Instead, environmental actions appear to bring these elements together (in line once again with Latour's notion of the hybrid). Within these particular re-constructions of the local and the global, there are also shifting constructions (or rather co-constructions, Irwin, 2001) of the social and the natural at work. In this way, the establishment of a dichotomy between the social and the ecological dimensions of these cases is as unhelpful as the adoption of a similarly bifurcated approach to the local and the global.

Fifthly, the two cases do suggest further questions of the relationship between local initiatives and democracy. Given the general advocacy of 'grass roots' initiatives within localism, these examples indicate, on the one hand, the practical implications (and potential success) of particular groups of citizens working towards sustainability but, on the other, the rather constrained (or perhaps re-defined) nature of the 'democracy' involved within these social and institutional processes. Broadly put, there is nothing self-evidently 'democratic' about the initiatives being discussed here – even if they do represent innovative social and institutional arrangements.

Sixthly, as to the wider applicability and relevance of success stories such as these, our answer – perhaps disappointingly to some – is that it depends on the robustness and imaginativeness of the actor networks involved. However, it does seem safe to say that no 'solution' (be it toilet or coolant) will work unless it embodies ideas that are held in common by the people for whom the object is intended. Thus, the separating toilet could emerge as a viable sanitation system for the villagers of Baalarna, because it 'fitted' well with the villagers conception of a good quality of life. In similar vein, the hydrocarbon coolants fitted well into Greenpeace's conceptual frame (i.e. limiting the emission of ozone depleting gases), which they extended to the consumers through their marketing activities and which offered the refrigeration industry sufficient proof of a potential market. The success of initiatives like the ones presented here lies less in the inherent properties of the technologies themselves and more in their presentation and in the way that they have been able to enrol other interested actors and give them sufficient legitimacy to 'survive'. This involves a multitude of actions that translate and stabilise heterogeneous elements into specific technologies and networks, i.e. making three-layered glass windows, insulation, and separating toilets essentials of an ecological village. These developments are enacted through the various actors' behaviour and their interpretation of other actors' response to their behaviours.

Reflecting generally on these cases, therefore, we can identify a number of emergent characteristics: the mutual embedding of the local and the global within particular initiatives, the analytical inadequacy of attempts to 'black box' or constrain the local (notably, in spatial terms), the collapse of a straightforward dichotomy between these two elements. Rather than seeking to categorise these cases as either 'local' or 'global', 'bottom-up' or 'top-down', 'centralised' or 'de-centralised' we have drawn attention to their characteristic complexity, inter-penetration and hybridity. In doing so, we have also raised a number of issues about the character of citizenship and social-natural relations in such cases.

These observations lead us also us to another general point. On the basis of the cases presented here, it does appear that categorisations into either the local or the global are very much human judgements rather than being straightforward and given characteristics of the cases themselves. Is Baarlana an example of the 'local' in action or of a national scheme coming to fruition? Is the Greenfreeze case simply another story of industrial corporations conducting an innovation process – or a local network establishing itself with global implications for sustainability? Of course, such points could be argued either way but our emphasis is on their social and institutional construction. They represent human categorisations from particular social viewpoints.

So what does this suggest for social scientific analysis and for policy engagement? Our intention is to encourage further discussion and debate. In that spirit, we would like to suggest the following:

- That talk of local initiatives and bottom-up strategies needs to be tempered with a more analytical understanding of the dynamic relationships involved;

- That rather than advocating either the local or the global, we should be addressing their re-combination and re-construction within particular social and ecological settings;
- That the language of 'local/decentralised/bottom-up' implies a spatial black-box which may not be justified;
- That the construction of such categories is a social and institutional process which itself requires further study and analysis (so that we should explore *localising* and *globalising* narratives rather than seeing such conventional categories as pre-determined or fixed).

In policy terms, our suggestion is that the increasingly commonplace rhetoric in favour of the 'local/de-centralised/bottom-up' be tempered with an awareness of the inadequacy of such spatial metaphors (since the local cannot adequately be defined in geographical terms alone) and the inseparability of local settings from wider relations of knowledge, expertise, institutional action, participation and innovation. This is not to deny the power of local rhetoric as a stimulus to action – nor the necessity of criticising and evaluating the operation of 'global' environmental institutions. However, there may be considerably greater potential within what we have termed 'local-global coalitions' than is generally recognised.

In more conceptual terms, the approach suggested here opens up many possibilities for further research and discussion. In particular, abandonment of the assumption that categorisations of the local and the global are intrinsically fixed and 'given' opens up a larger social scientific discussion of the construction of such entities within environmental debate. Rather than simply accepting such categorisations as unproblematic, our attention is drawn to the social and ecological significance of localising and globalising talk. Viewed in that way, we can move beyond the politics of localism and towards a consideration of the changing local-global coalitions being constructed within the environmental arena.

ACKNOWLEDGEMENTS

The authors are extremely grateful to all the researchers on the Social Management of Environmental Change project (Maj Andersen, Janet Rachel, Philip Vergragt, Hugo Verheul). We also thank the European Commission for its support under the programme of Research on Social and Economic Aspects of the Environment: DGXII/D/5.

REFERENCES

Andersen, M. (1995) Case report: Baalarna (Annex 1, case no. 18) in *The Social Management of Environmental Change - Final Report*, A. Irwin, S. Georg and P. Vergragt, European Commission.
Beck, U. (1992) *Risk Society: towards a new modernity*, London, Newbury Park, New Delhi: Sage.
Callon, M. (1986) The Sociology of an Actor-Network: The Case of the Electric Vehicle, in *Mapping the Dynamics of Science and Technology*, M. Callon, J. Law and A. Rip eds. London: Houndmills.
Czarniawska, B. (1997) *Narrating the Organization: Drama of Institutional Identity*, Chicago and London: The University of Chicago Press.
Donaldson, T. and L. Preston (1995) The Stakeholder Theory of the Corporation: Concepts, Evidence and Implications, *Academy of Management Review*, 20(1):65-91.
Dryzek, J. S. (1997) *The Politics of the Earth - Environmental Discourses*, Oxford: Oxford University Press.
Eder, K. (1996) *The Social Construction of Nature: a sociology of ecological enlightenment*, London, Thousand Oaks, New Delhi: Sage.
Freeman, R. E. (1984) *Strategic Management: A Stakeholder Approach*, Boston: Pitman.
Gabel, L. H. (1995) Environmental Management as a competitive strategy: the case of CFCs, in *Principles of Environmental and Resource Economics*, H. Folmer, H.L. Gabel and H. Opschoor eds. Cheltenham: Edward Elgar.
Haaland, T. (1994) Greenpeace Rapport [Report], no. 1.
Hannan, M. T. and J. Freeman (1989). *Organizational Ecology*, Cambridge, MA: Belknap Press.
Harrison, J. S. and R.E. Freeman (1999) Stakeholders, Social Responsibility and Performance: Empirical Evidence and Theoretical Perspectives, *Academy of Management Journal*, 42: 479-485.
Hartman, C. L. and E.R. Stafford (1998) Crafting 'enviropreneurial' value/chain strategies through green alliances, *Business Horizons*, 41(2):62-72.
Irwin, A., S. Georg and P. Vergragt (1994) The social management of environmental change, *Futures*, 26(3):323-34.
Irwin, A., S. Georg and P. Vergragt (1995) *The Social Management of Environmental Change - Final Report*, May 1995, European Commission.
Irwin, A. (1995) *Citizen Science - A Study of People, Expertise and Sustainable Development.* London: Routledge.
Irwin, A. (2001) *Sociology and the Environment: a critical introduction to society, nature and knowledge*, Cambridge: Polity.
Jacobs, M. (1997) The Quality of Life: Social Goods and the Politics of Consumption, in *Greening the Millenium - the new politics of the environment*, M. Jacobs ed. Oxford: Blackwell Publ. 47-61.
Jamison, A. and P. Ostby eds. (1997) *Public Participation and Sustainable Development: comparing European Experiences*, PESTO Papers 1. Aalborg University. Available from: Dept. of Development and Planning, Aalborg University, Fibigerstraede 13, DK-9220 Aalborg, Denmark.
Latour, B. (1993) *We have never been modern*, London: Harvester Wheatsheaf.
Law, J. (1992) Notes on the Theory of the Actor-Network: Ordering, Strategy and Heterogeneity, *Systems Practice*, 5(4):379-393.
Law, J. (2000) On the Subject of the Object: Narrative, Technology, and Interpellation, *Configurations*, 8 (1):1-29.

Lerner, S. (1997) *Eco-pioneers: Practical visionaries Solving Today's Environmental Problems*, Cambridge MA : MIT Press Books.

Long, F. J. and M.B. Arnold (1994) *The Power of Environmental Partnerships*, Washington DC: MEB/World Resources Institute.

Macnaghten, P. and J. Urry (1998) *Contested Natures*, London, Thousand Oaks, New Delhi: Sage.

Mander, J. and E. Goldsmith (1996) *The case against the global economy and for a turn toward the local*, San Francisco: Sierra Club Books.

Marvin, S. and S. Guy (1997) Creating myths rather than sustainability: the transition fallacies of the new localism, *Local Environment*, 2(3):311-8.

O'Riordan, T. and H. Voisey eds (1998) *The Politics of Agenda 21 in Europe*, London: Earthscan.

Szerszynski, B. (1996) On knowing what to do: environmentalism and the modern problematic, in *Risk, Environment and Modernity – towards a new ecology*, S. Lash, B. Szerszynski, B. Wynne eds. 104-37 London, Thousand Oaks, New Delhi: Sage.

Verheul, H. (1995) Case report: Greenfreeze (Annex 1, case no. 8) in Irwin et al. (May 1995)

Yearley, S. (1996) *Sociology, Environmentalism, Globalization: reinventing the globe*, London, Thousand Oaks, New Delhi: Sage.

Young, S. (1997) Local Agenda 21: The Renewal of Local Democracy? in: M. Jacobs (ed.) op cit 138-47.

Chapter 5

SUSTAINABLE BUSINESS

Learning-Action Networks as Organisational Assets[1]

NIGEL ROOME

SARAH CLARKE

1. INTRODUCTION

This chapter presents the results of a study of the *learning-action network* that influenced the environmental management and sustainable development practices in a Canadian company. This company was externally acknowledged as a leader of corporate environmental management practice throughout the 1990s. During the study, the company also began to distinguish between its commitment to corporate environmental management and the broader, emerging concept of corporate sustainable development. The thinking behind the research can be traced to the experiences of one of the authors in environmentally-driven, multi-party collaborative initiatives in the UK during the mid- to late-1980s (Vittery and Roome, 1989). These initiatives were developed by environmental non-government organisations (NGOs) as part of their policy response to the World Conservation Strategy (IUCN, 1980) which argued for new organisational forms that cut across existing sectors, functions and disciplinary boundaries. These and similar initiative are described by Carley and Christie (1993). Operationally these initiatives linked organisations that shared a physical space (e.g. river basin

[1]This chapter is a revised version of a paper originally published in *Business Strategy and the Environment* 8 (5): 296-310, Copyright John Wiley & Sons Ltd. and ERP Environment.

T.J.N.M. de Bruijn and A. Tukker (eds.), Partnership and Leadership, 77–101.

or industrial region) in new ways to address environmental concerns. These partnerships involved joint action to resolve problems inherent to systems (O'Riordan, 1971) or which arose from turbulence in organisational domains (Emery and Trist, 1965; 1973). They involved the establishment of collaborative structures to develop shared perceptions of problems and to determine and undertake agreed courses of action (Gray, 1989). Many of these initiatives involved companies, but they were invariable led by NGOs (Hartman and Stafford, 1998).

In the late 1980s, the shift in corporate environmental management in some companies from traditional non-strategic, reactive approaches to more strategic, proactive stances (Hunt and Auster, 1990), provided the possibility that companies might engage in multi-party collaborative initiatives as a way of linking their business and environmental concerns. From a sustainable development viewpoint this approach is advocated in the Brundtland Report (1987) and Agenda 21 (Earth Summit, 1992) and, the importance of inter-organisational and intra-organisational links and networks to corporate environmental management and sustainable development has been discussed in the literature (Roome, 1994; Clarke and Roome, 1995; Clarke 1998).

The central ideas that informed this study were that:

- Environmental management and sustainable development require companies to acquire knowledge that is not ordinarily found in their existing repertoire or experience.
- Environmental management and sustainable development require companies to participate in forms of collaborative action that link traditional business issues to a set of environmental and social concerns.
- The development of the knowledge and action required involves a broad set of actors with an interest in a company's activities including its strategy, technology management, environmental management and sustainable development.
- A multi-party mechanism to bring about the development of knowledge and provide the ground for collaborative action for environmental management and sustainable development is a learning-action network.

In this chapter, a learning-action network is defined 'as a set of relationships which lay over and complement formal organisational structures linking individuals together by the flow of knowledge, information, and ideas' (Clarke and Roome, 1995). These networks "are embedded in the complex of organisational and social relationships, management structures and processes that constitute business and its social context" (Clarke, 1999). Companies that seek to develop knowledge and action through learning-action networks which span their internal and external stakeholders have been termed 'meta-textual' organisations (Roome, 1997).

The research presented here is part of a larger study (Clarke, 1999) initiated in 1992/3 prior to the publication of much of the theoretical literature on corporate environmental and sustainable development. The literature focusing on multi-actor initiatives, compatible with our notion of the learning-action network, includes the multisystem perspective of ecologically sustainable organisations (Starik and Rands, 1995); strategic bridges between business and environmental organisations (Westley and Vredenburg, 1991; Sharma, et al., 1994; Stafford et al., 2000); institutional perspectives of inter-organisational learning (Jennings and Zandbergen, 1995); and adjustments in the structure of industrial networks in response to external environmental change (Östlund, 1994). More recently, Rowley (1997) has contributed to the discussion on stakeholder interactions by developing a stakeholder network theory based on the concepts of network density and the centrality of an organisation in its stakeholder network. Others, notably Hart (1995), have addressed the demands on organisational resources and capabilities arising from these multi-actor relationships.

The research design used in this study involved ethnographic methods. These permitted intimate, detailed observations of the interaction between the learning-action network that surrounds a company's formal approach to environmental management and sustainable development. The research used a grounded-theory approach; accordingly, it formulated propositions rather than tested hypotheses derived from existing theory (Strauss and Corbin, 1990).

The chapter is organised in four sections. The first section sets out the background to the study and the research methods. The second section describes the company and presents the findings of the case analysis. The third section discusses these findings in the light of previous and subsequent research. It also draws on the authors' emerging ideas about how corporate environmental management and sustainable development oblige the development of highly networked or meta-textual organisations as a basis for learning and action. It concludes by offering a set of propositions for future research.

2. BACKGROUND AND METHOD

This chapter examines how a publicly listed Canadian utility, known in this chapter as 'SysTec' (pseudonym) used networks of learning and action to define its approach to environmental management and sustainable development practice. It specifically focuses on how SysTec's internal strategic process and personnel supported the move toward an increasingly open approach to stakeholders by the corporation. The chapter analyses the characteristics of SysTec's approach and its ability to identify and act on elements of the environmental and sustainable development agenda. The case study formed part of a larger grounded theoretical study of the move toward more sustainable technology management in companies in Canada and the UK. The empirical work for this larger study was undertaken between 1993 and 1995. Aside from SysTec, the main project studied a learning-action network that surrounded technology management in two other companies, 'EngTec' and 'HiTec' (also pseudonyms). At the beginning of the study, SysTec was a subsidiary of EngTec. By the end of the study, EngTec had sold SysTec. A full case study of HiTec is found in Clarke (1998). This learning-action network is shown schematically in Figure 1. The present study reports on only SysTec and part of this learning-action network.

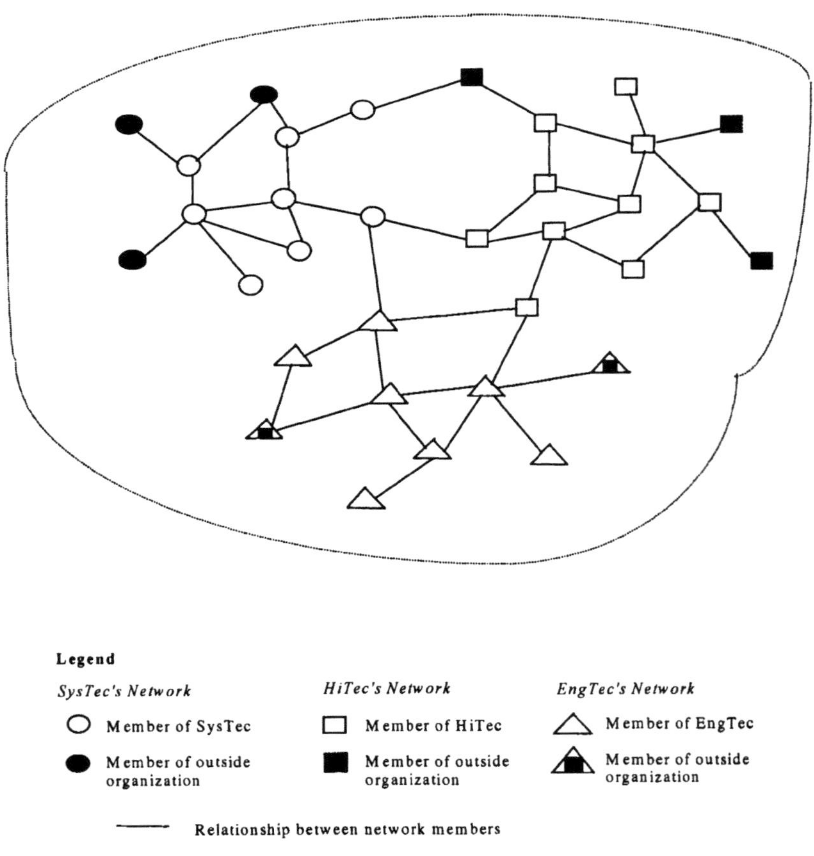

Figure 1. Schematic of the network as a whole

Figure 1 illustrates SysTec's linkages through third-party organisations, such as an industry association or direct relationships between employers of EngTec and SysTec. The focal companies were also connected through the products and services they each provided one another. This network was not apparent to its members. It was only revealed through the inquiry-process used in the research. The learning-action network is dynamic. It has 'fuzzy' boundaries and its overall shape changes as new needs arise and as responses are developed by network members. While some parts of the network were highly structured around existing organisational systems and relationships, other parts were very informal emerging out of personal contacts.

This study draws on interviews with more than 35 members of SysTec's learning-action network, identified using a 'snow-balling' technique (Jorgensen, 1989). These key informants included SysTec senior managers, environmental specialists and engineers. Additionally, members of the wider external network, identified by SysTec employees as influential in the changes taking place in the company, were included. Informants were asked to identify other individuals who were influential in shaping the ideas and actions that impacted the company's strategy, environmental performance and practices and technology (Clarke, 1997). Documentation available in the public domain was also drawn on in the research.

3. CASE-STUDY AND EMPIRICAL FINDINGS

This case study describes SysTec and the interaction between the company's strategic management processes, its environmental initiatives, and the learning-action network. The case study is set out in four parts. SysTec's business and its context are described. The way SysTec's approach to strategic management connected with the company's environmental thinking and practice is presented to highlight the role of interpersonal networks in the emergence of the company's environmental practice. SysTec's evolving response to the environment and sustainable development agendas are reviewed and finally, the implications for sustainable development are described.

3.1 SysTec's business and context

Founded over 150 years ago, SysTec's business is the distribution of a non-renewable resource. It operates in specific (franchise) areas, where it is responsible for storing, transmitting, distributing, selling and supplying this resource. It is a relatively small Canadian company with 3,000 employees, relying heavily on externally contracted research and development. The company is well known for its strong commitment to the communities it serves. Within the company there is a 'culture' of individual responsibility grounded in an ethic of corporate citizenship (SysTec's, *Annual Report*, 1993). This value underpins all actions of the company and the attitudes of its personnel contribute to an organisational philosophy to be a *role model for others* (SysTec's *Earth Day Annual Report*, 1993) and a belief that it is *part of the solution* for a more environmentally sustainable future (SysTec's *Our Resource and the Environment*, 1990).

SysTec operates in a regulated industry where the rates SysTec charges for its products and services and how it distributes its profits are controlled. The regulator (who weighs SysTec's business case against the general needs of the community and specific environmental interests) makes strict demands on the corporation's environmental initiatives and research programs. This regulatory regime means that any action, by other companies in SysTec's franchised area, to vary their rates, their community practices, or environmental interests can have implications for SysTec and its competitors. SysTec is acknowledged as a leader in environmental management. While the company has not purposefully sought this leadership position, companies voluntarily ask SysTec for advice and seek its leadership in environmental matters. In these ways, SysTec's activities can set standards for other companies.

SysTec shares ideas with other companies supplying the same resource in other franchise areas, particularly in relation to environmental concerns. This is done informally through one-on-one contacts and formally through research partnerships, industry association initiatives, and community-wide projects. Relationships with companies, which supply substitute non-renewable resources in SysTec's franchise area, are less harmonious.

3.2 Strategic planning, environmental management and sustainable development at SysTec

Prior to 1989, SysTec's environmental activities involved a loose set of initiatives required by legislation and regulation, voluntary initiatives originating from employee interests such as waste management and office recycling, and general 'housekeeping' initiatives and programs designed to improve operational efficiencies. During the late 1980s, SysTec encountered rising concern about environmental issues and demands from the general public and its regulators. These concerns included the environmental issues provoked by the Montreal Protocol on ozone depletion and the broader issue of global climate change. There were concerns about the need for better resource and product stewardship. These external pressures were supported by a pull from within the company, as employees sought to make a contribution to environmental improvement, consistent with the company's ethic of social responsibility.

SysTec was prompted to develop a more systematic approach to environmental concerns. In 1989 an Environmental Strategy Committee (ESC) was formed (comprising senior managers from across the company, including its environment department) and a process of organisational change was begun. This resulted in the formulation of an environmental management system (EMS) which, by 1994, was an integral part of the company's strategic decision making processes.

At this time SysTec's approach to environmental management was underpinned by four principles: (1) it was 'low key'; (2) it led by example; (3) it worked with stakeholders; and (4) the approach was strategically oriented. The 'low key' approach meant the company did not seek highly visible recognition of its environmental programs or community projects. For example, one environmental manager noted:

> "I think because we're regulated we can't throw gobs of money at it [community projects]. Our level of sponsorship is basically $5,000 or less...And basically our demands for recognition are usually low key and subtle...There is a recognition [in the Environmental Services Department] that we try to do the right thing for its own value, not necessarily because we see it as part of a larger marketing strategy...We don't necessarily need to be broadcast to everyone as saviours of the world." (SysTec Interview III).

SysTec 'led by example,' seeking to put its own house in order (SysTec's *Environmental Plan*, 1994:1) and by acting on national and international environmental initiatives, such as climate change. The rationale behind this approach is that:

> "If you say that you are part of the environmental solution...that implies an obligation that...your own operations will be beyond reproach. But to me it also implies an obligation that you will also help develop what is a solution to the concerns before [you] and you can't do that alone, you have to do that as part of a larger society." (SysTec, Interview III).

Partnerships and stakeholder involvement were important aspects of SysTec's environmental philosophy. The company emphasised the need for improved mechanisms for communication with stakeholder groups and placed importance on education about environmental issues for their employees and for the general public. SysTec sought to understand the perspectives of different stakeholders rather than seeking to influence or dominate their opinion. It used insights from these interactions to review and improve the company's environmental programs and activities.

> "...[when] we develop partnerships [we develop] the ability to be able to sit down and listen to these people and get a better understanding of what their interest is, their perspective and, in terms of a company doing its strategic planning,...how it should approach the environmental issues, that kind of grass roots polling as it were, building trust..." (SysTec, Interview III).

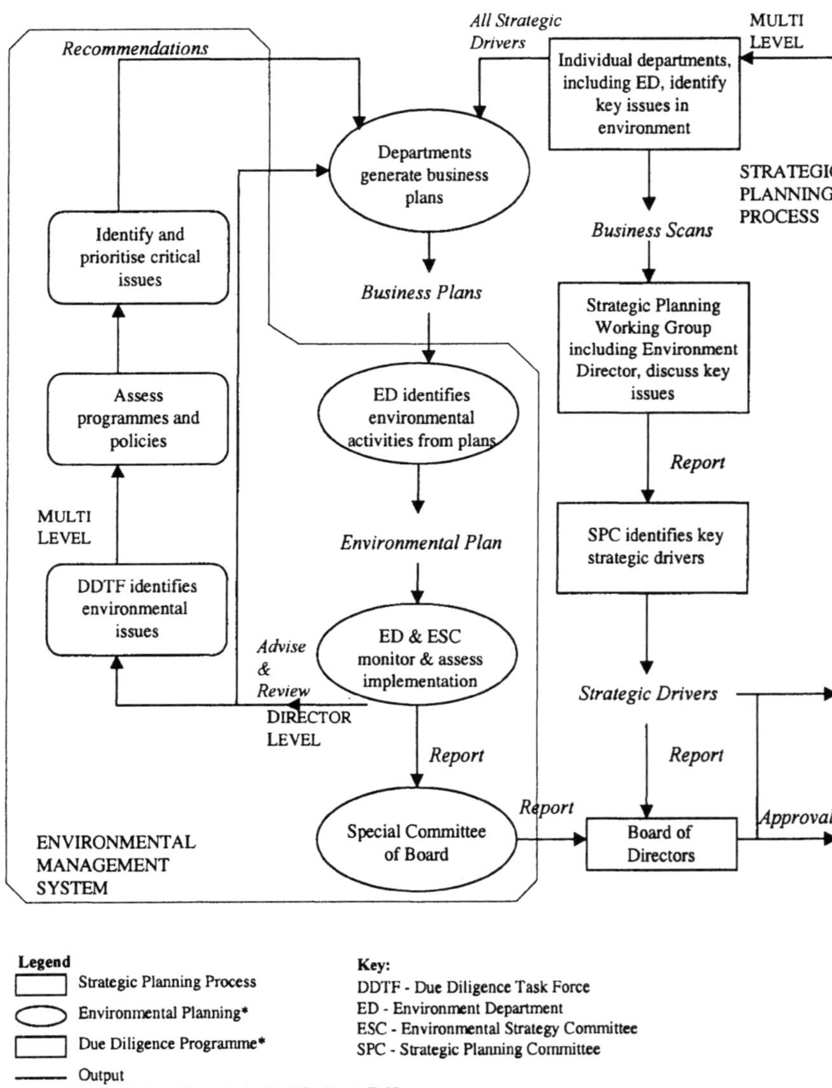

Figure 2. Relationship between strategic planning process and EMS at SysTec

SysTec's strategic planning and decision making was also key to its ability to respond to changing circumstances in the business environment. It enabled the company to identify issues in the business environment, to assess their implications, and to initiate and implement programs in response. This supported the rapid changes made by the company in its environmental activities. SysTec's strategic management process led to 'continual improvement' in its environmental programs, through identification of new environmental concerns, implementation of new environmental activities, and revision of existing approaches.

SysTec's strategic planning process was designed to identify factors in the business environment that would shape its activities in the coming year. These factors are described as *key strategic drivers* (SysTec, Interview I). The strategic planning process predated the emergence of environment and sustainable development on SysTec's corporate agenda. Since 1989, successive 'environmental' issues were identified by this planning cycle. A key part of the strategic organisational response to these environmental concerns was to integrate the company's EMS and the Due Diligence Program (DDP), required by law, into its planning process.

The description provided by informants of SysTec's annual strategic planning process, in operation after 1994, is represented diagrammatically in Figure 2. The figure shows the way the EMS and DDP linked to the overall process of SysTec's business planning cycle.

All departmental directors were required to produce a *scan of the business environment [context]* to identify issues of strategic importance to their department. Directors discussed these business scans in the company's Strategic Planning Working Group (SPWG) and more informally through their day to day operational relationships. This was designed to make sure that each department was aware of the complete range of factors that were affecting all SysTec's departments. A team of Vice Presidents (VP) and senior managers, members of the Strategic Planning Committee (SPC), reviewed these business scans and this committee decided which strategic drivers would guide the company's activities for the year. In practice, several departments identified environmental issues as critical business concerns.

A due Diligence Task Force (DDTF), drawn from managers across the company, was responsible for the DDP process. It considers the company's operations and areas of potential environmental risk, determining whether SysTec's procedures are adequate or need improvement. DDP aims to keep abreast of activities in the industry as a whole:

> "not necessarily to be ahead of the pack, but certainly that [they've] got a plan in place to handle it, just in case something happens" (SysTec Interview II).

The due diligence exercise takes place before the business planning cycle begins. In this way departments can:

> "[R]ead it, assimilate it, ask questions and then sit down to decide what should I and my department be doing in terms of cost effectiveness, productivity, environmental protection, customer service?" (SysTec, Interview III).

The DDP continuously assesses and integrates environmental risks into the business planning cycle, even if environmental issues are not specific strategic drivers.

The environmental planning and DDP loops shown in Figure 2 form the basis of the company's EMS with the strategic planning process enveloping the company's EMS. The three management processes, seen in Figure 2, facilitated the integration of environmental concerns and responses into all levels of SysTec's decision making, from strategic to operational. As one environmental manager described it:

> "[W]e try to use existing business planning procedures, so we don't reinvent the wheel, but we try to embellish and enhance the existing ones so that they have the environmental factor...at the end of the strategic planning process...[senior managers] come up with the strategic themes that they think are important, [that] the business managers should think of. In the last few years, environment has been one of those. Now it doesn't have to be there all the time...But at least it gets fair weight. [The Environment Director] is very much involved in that process. So it's an opportunity to directly make presentations as to what environmental issues are before us, pressures, opportunities, threats that we should factor in." (SysTec, Interview III).

Business departments bring forward their business plans once strategic drivers are established through this process. These plans detail the environmental activities guided by the company's DDP. Members of the environment department compile the company's annual environmental plan from these business plan initiatives, assist in program development and monitor and review the company's environmental performance. The Environmental Director is responsible for reporting progress to a Special Committee of the Board of Directors, which reports to the full Board (SysTec's *Annual Report*, 1991:13).

The environment emerged as a strategic driver for the corporation in 1991 and 1992. It fell off the list of drivers in 1993, as the initial response to the 1991 and 1992 agenda were put into operation. Several key informants referred to this as the process through which environment became embedded in the company's culture and operational activities. Although the environment was not a strategic driver in 1993, it was identified as an important company performance measure. Consequently, departments began to include environmental performance in their business plans to measure success in meeting the environmental commitments set out in the company's environmental policy and environmental principles. In 1994, environment again emerged as a strategic issue but under the umbrella of corporate responsibility and citizenship. Other strategic drivers included employee development, market share and growth targets and financial goals.

In 1994, the ESC established a set of environmental priorities for the company. These ranged from low priority actions, such as waste management and environmental manuals, to high priority actions, for example environmental training and government and environmental partnerships. These action areas led to the development of an enhanced set of performance objectives with targets and measures designed to ensure that the company continuously improves its environmental performance (SysTec's, *Environmental Plan*, 1994:23-24).

3.3 SysTec's emerging approach to sustainable development

SysTec's recent commitment to the environment focuses around the values and goals of the ethic of corporate responsibility that has guided the company throughout its history. These emphasise sensitivity to the needs of customers and employees, to the concerns of the local communities in which it operates and, more recently, to the environmental implications of its business. In its *Annual Report*, the company acknowledges its responsibility to enhance shareholder value, but sees this as only one of its goals. SysTec's environmental programs are also influenced by the regulatory circumstance under which it operates, with its expectation that it will continue to promote new and improved approaches to environmental protection and management.

SysTec's environmental position is shaped by the environmental impact of the resource it supplies. The company is conscious of health and safety responsibilities around the efficient and safe use of this resource. SysTec, together with other members of its industry, views its product as part of today's solution to society's search for a more sustainable future. The product is widely acknowledged as having significantly less environmental impacts than commercially available substitutes. In company literature, SysTec promotes its resource as not perfect but a great step in the right direction (SysTec's *Our Resource and the Environment*, 1990) while acknowledging that stewardship of the environment is everyone's responsibility (SysTec's *Environmental Initiatives*, 1991). SysTec claims to be 'part of the solution' to a more environmentally sustainable future (SysTec's *Our Resource and the Environment*, 1990).

SysTec was an early signatory to the International Chamber of Commerce's (ICC) Business Charter for Sustainable Development. According to SysTec: *The Charter comprises sixteen principles for environmental management which, for business, are vitally important aspects of sustainable development* (SysTec's *Environmental Plan*, 1994:2). However, apart from this commitment, and a passing reference to sustainable development under the company's list of strategic priorities in 1994 (SysTec's *Environmental Plan*, 1994:23), there are no explicit references to sustainable development in the company's published documentation. SysTec does not define sustainable development or provide an explanation of its implications for the company.

SysTec's approach was initially consistent with the emphasis in the ICC charter on the management of environmental effects and the notion that its overall business provides a bridge from an unsustainable past to a more sustainable future. However, SysTec's environmental management approach has become more sophisticated and increasingly consistent with the broader notion of sustainable development. SysTec's approach to environmental policy in 1989 (SysTec's *Statement of Environmental Principles*, 1994) emphasised a commitment to conduct operations in an environmentally sensitive manner, promote employee and public awareness of environmental issues, encourage use of its product as the environmentally preferred choice, and develop technology to improve efficiency in resource utilisation. By 1994 SysTec's *Statement of Environmental Principles* read:

> "[SysTec] recognises the intrinsic value of nature and is committed to conducting all of its operations in an environmentally responsible manner, with a view to protect and maintain the environment for future generations. This Statement of Environmental Principles guides the Company in achieving its commitment to environmental protection and citizenship. It is the obligation of every employee to understand his or her environmental responsibilities..."

This document does not specifically reference sustainable development but the explicit recognition of the *intrinsic value of nature* and *citizenship* together with the company's commitment to environmental stewardship, partnership, and learning with others suggests an approach that goes beyond conventional corporate environmental management. It certainly represents a step beyond the company's earlier identification of itself as *a responsible corporate guardian of a non-renewable resource* (SysTec's *Environmental Plan*, 1991). The emphasis on understanding different stakeholder values, sharing ideas and working co-operatively with different groups implies an embedded understanding of the value to the company of engaging in relationships that bring about organisational and social learning and change.

3.4 Characteristics of SysTec's approach to more sustainable practices

Some of the key aspects of SysTec's approach to environmental management and sustainable development are described next. Table 1 pinpoints major milestones in SysTec's environmental management approach during the critical period between 1989 and 1994. Table 1 also identifies the process of change and the outcomes that are linked to SysTec's learning about how to be more sustainable. These changes can be categorised as: management structures and processes; communication activities; partnership and collaborative initiatives. Each of these is discussed below.

Table 1. Major Environmental Milestones at SysTec

Period	Philosophy	Major Milestones
1989	Environmental Responsibility and Protection	• Environmental Strategy Committee formed • Environmental Policy endorsed • Annual environmental planning process initiated and first Environmental Plan produced
1990-1993	From Environmental Protection to Environmental Stewardship	• Environmental Services Department established • Code of Environmental Ethics added to Policy • Corporate environmental logo launched and a set of informational pamphlets produced about global environmental issues and SysTec's responses to these concerns • Environmental Management System (EMS) and Due Diligence Program (DDP) introduced • Environmental audit process initiated • Waste Management target surpassed and redefined • Partnerships established, interest group meetings held, environmental research collaborations initiated • Collaborative projects developed in response to regulatory requirements • Earth Day Committee set up and annual Earth Day activities launched
1994	Environmental Stewardship and Citizenship	• Statement of Environmental Principles replaced Code of Environmental Ethics • Environmental Plan revised as a more strategic document • EMS and DDP strengthened and independent assessment of environmental activities commissioned • Environmental priorities established and additional environmental targets set • Comprehensive training program launched • Partnership and collaborative activities continued

Management structures and processes: As detailed earlier, the strategic planning process, EMS, and environmental committee structure identified the environment as a strategic driver and ensured that all levels of personnel were involved in the development of environmental programs within the company. A company-wide training program was begun in 1994, to ensure that the company's environmental commitment was understood, with roles and responsibilities clearly identified. This training program complemented the management process. In addition, the company made environmental information available through an in-house electronic bulletin board covering environmental issues, environmental displays and publications.

SysTec suggests that the strategic process incorporated the new environmental management and sustainable development agenda in a number of ways. This involved:

1. problem identification - where environment was picked up as a strategic driver;
2. operationalisation - where environment was integrated into existing activities with new systems adopted and routinised, including additions to the strategic planning process itself; and
3. redefinition - where environment was dropped as a stand-alone strategic driver to re-emerge under the umbrella of a new strategic driver described as 'Corporate Responsibility and Citizenship'.

The ability of the strategic process to recognise issues and to create the ground for integration is seen as important.

Integration took place on a number of levels. First, separate strands of the 'environmental agenda' before 1989 were integrated into a single strategic issue at the beginning of the 1990s and an overall organisational response was formulated to the environmental agenda as a whole. By the end of the case study, the newly fashioned strategic term 'environment' was recast and integrated as part of a wider issue connected to corporate social relationship. This helped SysTec move its conception of its environmental impacts and activities from environmental protection and management toward a broader notion of sustainable development. Second, SysTec learned how to integrate these issues within its organisational structure and its strategic and business planning processes that informed organisational change. Finally, the overall 'identity' of the organisation, the connection between its culture, values and strategic process, technology and resource base, was able to accept the new ideas about environment and sustainable development. In particular, this is seen through the company's predisposition to communication and partnerships.

Communication activities were important as SysTec sought to share ideas with different internal and external stakeholders and gain an understanding of the views of others. SysTec published educational leaflets on specific environmental issues, which affected the company and its industry in general, for example, ozone depletion and global warming. This was combined with public information sessions and displays together with environmentally related community projects to communicate the company's position and to raise awareness about environmental concerns.

The company's belief that it was 'part of the solution' to environmental problems meant it did not feel defensive about its environmental position and ensured that all sides of an environmental issue were addressed. Various environmental NGOs were encouraged to meet with the company to discuss their concerns. The purpose of these meetings was not to give "a 'Dog and Pony show' where we would tell then how wonderful we were and what we were doing" (SysTec, Interview I). Instead, environmental activists were invited to meet with a number of departmental directors to discuss what they were planning, their strategies for the future and how they saw the company's position developing. Before adopting this approach SysTec's senior management, with the exception of environmental managers, had only met interest groups in adversarial settings, such as at regulatory hearings. The environment managers played an important role in creating these settings and establishing the trust that enabled representatives of the stakeholder groups to meet with departmental directors. One environmental manager said that of these meetings:

> "It provides an opportunity for an outside group, not necessarily to see us as a black box but to see other places. It gives an opportunity for a lot of other people in...[SysTec] to hear first hand what a group is, instead of having it distilled through me. It gives them a chance to raise their own questions as well..." (SysTec, Interview III).

This same manager described it as a mutual learning process where each side was able to explain their position. Critically he noted:

> "There'd be areas where we agree to agree and areas where we agree to disagree and areas where we have opportunities to work together and areas were we need to work together towards resolving our differences."

These meetings helped to replace the stereotypes each side held about the other. These stereotypes portrayed business managers as "older men, with well pronounced bellies, smoking cigars", with environmentalists characterised as "former hippies with torn blue jeans" (SysTec, Interview III). Outcomes of these meetings were that the company developed a better understanding of interest groups' expectations and a clearer picture of what stakeholders believed SysTec should be doing, if it was to meet its claim to be part of the solution to society's environmental problems. Moreover, SysTec gained insights into the philosophy and tactics adopted by environmental groups.

These links reinforced SysTec's commitment to environmental concerns, enabling a wide range of external perspectives and values to be considered in their response to those concerns. This was strengthened by the company's emphasis on partnerships and collaborative initiatives.

Partnerships and collaborative initiatives: SysTec actively engaged in collaborative initiatives with many stakeholders: the general public, interest groups, industry members and industry associations, research associations and universities, regulatory and governmental agencies and competitors. The rationale for these partnerships was the realisation that the days of 'command and implement' management styles were over (SysTec, Interview III). SysTec increasingly recognised the need to find alternative ways to resolve issues of conflict.

Through more inclusive styles of management, they were able to gain a better understanding of the expectations of their customers, governments, regulators and special interest groups.

Aside from holding dialogue sessions with different interest groups, SysTec sponsored small environmental projects, such as the local zoo's 'adopt a pond program' as part of its community activities. SysTec also became an active supporter of Earth Day to educate company employees and customers about its commitment to environmental protection.

SysTec used partnerships to promote the use of its products as the environmentally preferred choice and to develop new technology that would support this goal. SysTec had a relatively small Technology and Development (T&D) Department, which focused on the development aspects of the continuum from 'Basic Research - Product Development'. SysTec developed the capability to leverage its basic research needs through partnerships with external research organisations. Moreover, part of T&D's charge was to modify the products used by the company's customers. It did this by encouraging manufacturers to undertake technology development, to fill technological gaps and bottlenecks. If the technological base was missing, it promoted research to fill the gap. SysTec was involved in a large number of research collaborations with local, national and international industry associations, research organisations and universities. It was involved in projects that spanned a variety of issues from industry specific technological concerns about operational equipment and their environmental implications, to more generic global environmental issues, such as climate change, involving basic research in, for example, atmospheric chemistry.

SysTec was active in persuading other industry players to have greater involvement in environmentally related projects outside the industry's 'traditional' areas of research. The company spearheaded several industry association collaborative initiatives, such as a handbook of environmentally related, industry-wide research, an annual conference on environmental issues affecting the industry and environmental training. Through its committee memberships SysTec was a key contributor in the formulation of an industry level environmental research policy. Senior personnel were assigned to multi-stakeholder collaborative initiatives, for example, a scheme to 'green' the city in which it had its headquarters, an initiative that involved representatives of environmental groups, engineering firms, government, management consultancies and business.

SysTec's strong relationships with its regulator as well as government departments and agencies at municipal, provincial and national levels were important. The regulator required SysTec to take part in two collaborative initiatives. One explored ways of internalising environmental costs in the industry, the other encouraged demand-side management programs as a means to minimise environmental impacts. These initiatives involved multi-stakeholder consultative and collaborative groups, which brought competitors, environmental interest groups, consumer groups and governmental representatives together to address their fundamental differences. This experience helped senior managers become aware of the need to address SysTec's approach to stakeholders, as they saw that they could not force these processes, the details of the agenda or the outcomes that emerged. Instead they saw the need to build consensus around shared understandings, with environmental groups demanding greater, more radical, action than most of the participating companies were prepared to undertake. SysTec was among the most progressive companies in these collaborations.

4. DISCUSSION

The findings from the study provide strong evidence of the importance of learning-action networks in the acquisition of new knowledge and the development of collaborative structures to bring about corporate environmental management and more sustainable forms of business activity. However, before discussing the characteristics of SysTec's approach to the development and acquisition of new environmental knowledge, it is important to address the overall embedding of environmental management thinking and practice in SysTec's existing processes as it adjusted to new environmental concerns and realities. It is also important to draw attention to the overall alignment between SysTec's values, strategic process and managerial attitudes, which provided organisational pre-conditions that influenced, and were influenced by, its emerging practice of environmental management. For example, SysTec operated around values that related to corporate citizenship. These were widely accepted in the company. The company's open, participative strategic processes enabled environmental issues to be identified and addressed, and then incorporated within that process. These newly adopted environmental practices were informed by principles, such as environmental stewardship and responsibility and by openness to the ideas and contributions of external stakeholders. These environmental management principles and processes were consistent with the pre-conditions at SysTec.

SysTec engaged in mutually iterative learning and change. For example, its organisational procedures were adapted as a result of new environmental management ideas. Environmental management ideas were then changed in light of these, now modified, organisational procedures. The openness to ideas within the company easily permitted this learning and change to take place.

SysTec's strong commitment to learning, its specific corporate values and its process of strategic planning meant that its was readily able to undertake change, and to incorporate environmental management into its organisational routines. Many companies appear to have found this difficult. Yet, SysTec rapidly developed new knowledge not found in its existing repertoires or experience. In other words, SysTec was receptive to, and capable of, acquiring this new knowledge.

In particular, SysTec's managers' abilities to work with a range of stakeholders as part of its traditional way of operating in T&D or the supply-chain was simply extended to collaboration with new stakeholders, even those with non-business interests. Although the partners with whom SysTec sought to learn and take action were new to most of the managers in the company SysTec set about trying to understand their needs and interests. It did not need to learn new ways to learn with them, it only had to adapt its established ways of learning to working with these new partners.

Environmental managers were specifically responsible for bringing members of their wider networks of environmental interests into contact with a number of senior corporate managers and department heads. Rather than retaining these network contacts exclusively to themselves, the environmental managers sought to spread the knowledge from their networks widely into the internal corporate structure of SysTec.

These observations can be related to the literature on strategic bridging (Westley and Vredenburg, 1991; Sharma, et al., 1994; Stafford et al., 2000) stakeholder networks (Rowley, 1997) and organisational resources (Hart,

1995; 1997). In the case of strategic bridges, as mechanisms that enable the exchange of environmental knowledge between organisations such as environmental groups and companies, the study of SysTec indicates the importance of establishing multiple bridges between people who work in different organisations. The metaphor of the strategic bridge, therefore, does not appear totally appropriate to the relationships found at SysTec. These are better represented as a rather fluid, network-like structure that involves learning and action by many people in the company and by many people and organisations in a company's 'stakeholder field'. SysTec was not involved in establishing a single 'strategic bridge'. It was engaged in developing multiple points of contact for learning and action. This fluid 'strategic network' provided the ground within which more formal 'strategic collaborative bridges' could arise between stakeholders and different parts of the company. This implies that SysTec was simultaneously involved in the development of strategic bridges, as formal collaborations, and in advancing a less formal network of interactions between the company and its stakeholders, of the kind envisioned by Rowley (1997). However, in contrast to the emphasis Rowley places on the concepts of network density and organisational centrality, the experience at SysTec reveals the importance of organisational pre-conditions and interpersonal capabilities in developing the potential of the learning-action network as an organisational asset.

The study identifies the crucial role of environmental managers in facilitating the process of network development and as the architects of multiple bridges, with collaborations then supported by other senior managers. The organisation's acceptance and support for the process of bridge building aided these developments. This support, or legitimacy, derived from SysTec's longstanding values and approach to collaboration, which enabled SysTec's environmental managers to develop the space to broker new relationships and build new bridges. These then helped inform organisational processes such as strategic planning or T&D management.

The study of SysTec reinforces Roome's (1994) contention that environmental management and the environmental dimension of technology development require new, more open, mechanisms for learning and collaboration and learning through collaboration. It extends Hart's (1995) proposition that capabilities, in drawing external stakeholders into dialogue with the company, are an important asset for companies involved in environmental management, but, it contradicts his later suggestion that these mechanisms should be used to encourage learning by stakeholders (Hart, 1997). Overall, the experience at SysTec identifies serious limitations with this view.

SysTec's experience shows that networks are not just for communicating a set of company messages to stakeholders. Instead they are a mechanism for mutual learning and action by the company (and its managers) and its stakeholders (organisations and individuals). This implies that learning is two-way. For example, stakeholder inputs could lead to the revision of SysTec's corporate strategy or to the establishment of new multi-party business and environment initiatives. In this sense, SysTec's strategy did not define the nature and direction of collaboration with the company's network of stakeholders. As a consequence of its company culture and attitudes SysTec's organisational boundary was porous to stakeholder issues and concerns and its strategy was flexible and responsive to these issues and interests.

The study suggests that a complex process of negotiation, learning, action and change between the company and its stakeholders was taking place to integrate environmental ideas into SysTec and to progress sustainable development. This involved the managers of the company and its dynamic network of stakeholders in a process of continuous mutual adjustment.

This process of learning and action is not simple. SysTec's networks with stakeholders were characteristic of its industry, technologies, organisation, personnel, culture and history. Not only are these networks dynamic, they contain a diverse array of values, experiences, interests and ideas within which a myriad of contested interpretations of environmental and sustainable development concepts are played and acted out (Myerson and Rydin, 1996; Clarke, 1997). Often these differences are found in the specialised use of language within different (stakeholder) communities (Roome, 1997). This implies that those responsible for facilitating learning, action and change through networks must have the capability to bridge these different communities or islands of knowledge and language (Clarke, 1997). The capability of these boundary-spanners (Mylonadis, 1993) or network champions (Roome, 1997) or catalysts (Clarke, 1999) to promote learning, action and change through networks appears a critical, yet under-researched, aspect of environmental management and sustainable development practice. It is suggested that during the period of the case SysTec assumed the character of a meta-textual, highly networked organisation, engaged in the development of knowledge and collaborative action through interaction with many internal and external stakeholders as a way to bring about environmental management and sustainable development.

5. CONCLUSIONS

A number of conclusions are drawn from the case study. SysTec's acknowledged leadership in environmental management is traced to the context, business and established values that provide it with a set of organisational pre-conditions. The regulatory setting and the Canadian approach to multi-party dialogue create a context in which it is legitimate for businesses to forward environmental responsibilities. The resources and technologies at the core of SysTec's business enabled it to position itself as environmentally friendly.

Proposition 1
Corporate effectiveness in developing responses to environmental concerns and sustainable development is critically influenced by the institutional context and organisational pre-conditions that pre-dispose an organisation to mutual learning and change with other stakeholder groups.

SysTec then drew on a number of internal pre-conditions that enabled it to learn and change to develop its environmental management capability and sustainable development potential. Important pre-conditions include the values by which the company operates and its strategic processes, which identified environmental issues as a future strategic driver. This process was open, participative and flexible enough to adapt the organisation to these concerns. SysTec's strategic planning process can be viewed as an organisational spiral, collecting issues from across the company, rather than the more narrowly defined, planned, hierarchical, closed process used by strategic planning groups in many organisations.

Proposition 2
Companies that are open and responsive to multiple perspectives are more disposed to acquire new knowledge and take actions that meet environmental management and sustainable development needs than those that develop knowledge and act within their existing resources.

The internal process at SysTec supported increasingly close relationships with its customers, its regulators and the communities in which it operated. This was supported by SysTec's willingness to be open, to view communication as listening as well as talking, to engage in many different learning opportunities, to act together with a wide range of stakeholders and to broaden their points of contact within the company. SysTec was able to make effective use of a rich network of knowledge.

Proposition 3
Companies that acquire knowledge which contributes to effective environmental management and sustainable development have access to managers with highly developed networks, networking skills and capabilities in facilitating change through those networks.

SysTec's environmental managers played an important role in facilitating settings in which learning was developed and translated into joint actions. They effectively bridged internal and external networks.

Proposition 4
Effective environmental management and sustainable development require companies to use networks of stakeholders as a means to inform, as well as confirm and validate their approach to environmental management or sustainable development.

SysTec developed a learning-action network form of organisation. This provided a frame or sounding board that influenced the company's ability to develop novel ways to meet the challenges it confronted in the economy, environment and society. Relevant networks provided feedback that enabled SysTec continuously to re-define the problems of environment and sustainable development and to shape and reshape its responses to old and new issues. It continues to provide opportunities for SysTec to act with other stakeholders to address fundamental changes in the socio-technical context of which SysTec is part.

Proposition 5
Effective environmental management and sustainable development involve inclusive networks for learning and action. The more inclusive a network is, the greater the demands on the 'process skills' of managers to reconcile the problems that stem from the difference of perspective and language used by network members.

Despite SysTec's acknowledged success in shaping its response to environment and sustainable development within its business, a number of issues remain. For example, there are questions about the extent to which SysTec was able to respond to the more far reaching demands of stakeholders network and to develop collaborative initiatives and projects while respecting fundamental differences of perspective that could not yet be bridged. SysTec was also aware that it had to learn, act and negotiate with many different stakeholders, yet it had begun to encounter problems; in particular, SysTec's managers encountered contested meanings and values and problems over the different uses of language in the learning-action network. These problems were beginning to influence the implementation of joint projects.

Proposition 6
Effective environmental management and sustainable development involve highly developed skills in facilitating inputs from multi-stakeholder networks at all levels of a company - strategic, environmental and technological as well as operational.

There was evidence that SysTec's competitors were beginning to imitate SysTec's programs and activities rather than undertaking change in their organisational culture, attitudes and processes to bring them into line with the pre-conditions for SysTec's success. The case study suggests the need for all the strategic aspects of a company to be open to new knowledge and new opportunities for collaboration through the mechanism of learning-action networks. The challenge for corporate strategists, technology managers, and environmental managers is to synchronise the 'soft inputs' from learning-action networks with more analytical skills of strategy formulation or technology development and management.

ACKNOWLEDGEMENT

The authors would like to thank the Erivan K. Haub Program in Business and Environment, Schulich School of Business, York University, Canada for its financial support of this research. The accuracy of the content of this chapter remains the responsibility of the authors.

REFERENCES

Brundtland, G-H. (1987) *Our Common Future* – Report of the World Commission on Environment and development, Oxford: Oxford University Press.
Carley, M. and I. Christie (1993) *Managing Sustainable Development*, Minneapolis MN: University of Minnesota Press.
Clarke, S.F. (1997) *Unravelling networks of learning in the search for more sustainable technology management.* Greening of Industry Network Conference, Santa Barbara, CA. 1997.
Clarke, S.F. (1998) Sustainable technology management: The role of networks of learning, in *Sustainability Strategies for Industry* N. Roome ed. Washington D.C.: Island Press.

Clarke, S.F. (1999) *Knowledge networks and sustainable technology management: An international study of organisational practice*, doctoral thesis, Ontario: Schulich School of Business, York University.

Clarke, S.F. and N. Roome (1995) Managing for environmentally sensitive technology: Networks for collaboration and learning, *Technology Assessment and Strategic Management*, 7, (2): 191-215.

Earth Summit (1992) *Earth Summit: United Nations Conference on Environment and Development*, London: Regency Press Corporation.

Emery, F. and E. Trist (1965) The causal texture of organisational environments, *Human Relations*, 18: 21-32.

Emery, F. and E. Trist (1973) *Towards a Social Ecology*, New York: Plenum Press.

Gray, B. (1989) *Collaborating: Finding Common Ground for Multiparty Problems*, San Francisco: Jossey-Bass.

Hart, S. (1995) A natural-resource based view of the firm, *Academy of Management Review*, 20, (4): 986-1014.

Hart, S. (1997) Beyond greening: strategies for a sustainable world, *Harvard Business Review*, January - February, 67-76.

Hartman, C.L. and E.R. Stafford (1998) Crafting 'environmental' value chain strategies through green alliances, *Business Horizons*, 41, (2): 62-72.

Hunt, C. and E. Auster (1990) Pro-active environmental management: avoiding the toxic trap, *Sloan Management Review*, 31, (2): 7-18.

IUCN (1980) *World Conservation Strategy: Living Resource Conservation for Sustainable Development*, IUCN-UNEP-WWF: Gland.

Jennings, P. and P. Zandbergen (1995) Ecologically sustainable organisations: an institutional approach, *Academy of Management Review*, 20, (4): 1015-1052.

Jorgensen, D. (1989) *Participant Observation: A Methodology for Human Studies*, London: Sage.

Myerson, G. and Y. Rydin (1996) *The Language of Environment: A New Rhetoric*, London: UCL Press.

Mylonadis, Y. (1993) *The Green Challenge to the Industrial Enterprise mindset: Survival Threat or Strategic Opportunity?* doctoral thesis, Sloan School of Management, MIT.

O'Riordan, T. (1971) *Perspectives on Resource Management*, London: Pion.

Östlund S. (1994) The limits and possibilities in designing the environmentally sustainable firm. *Business Strategy and the Environment*, 3, (2): 21-33.

Roome, N. (1994) Business strategy, R&D management and environmental imperatives. *R&D Management*, 24, (1): 65-82.

Roome, N. (1997) *The role of meaning, communities of knowing and action learning in the integration of business and the environment - a theory of action learning networks*, Greening of Industry Network Conference, Santa Barbara, CA, 1997.

Rowley, T. (1997) Moving beyond dyadic ties: A network theory of stakeholder influences. *Academy of Management Review*, 22, (4): 887-910.

Sharma, S., H. Vredenburg and F. Westley (1994) Strategic bridging: a role for the multinational corporation in third world development, *Journal of Applied Behavioural Science*, 30, (4): 458-476.

Stafford, E.R., M.J. Polonsky and C.L. Hartman (2000) Environmental NGO-business collaboration and strategic bridging: a case analysis of the Greenpeace-Foron alliance, *Business Strategy and the Environment*, 9, (2): 122-135.

Starik, M. and G. Rands (1995) Weaving and integrated web: multilevel and multisystem perspectives of ecologically sustainable organisations, *Academy of Management Review*, 20, (4): 908-935.

Strauss, A. and J. Corbin (1990) *Basics of Qualitative Research: Grounded theory procedures and Techniques*, Newbury Park CA.: Sage.

Vittery, A. and N. Roome eds. (1989) Partnership Review, Nature Conservancy Council & Shell (UK), Peterborough.

Westley, F. and H. Vredenburg (1991) Strategic bridging: the collaboration between environmentalists and business in the marketing of green products, *Journal of Applied Behavioural Science*, 27, (1): 65-90.

Chapter 6

ORGANISING INTERFIRM LEARNING

As the Market Begins to Turn Green

MAJ MUNCH ANDERSEN

1. INTRODUCTION

The recent rise of greening as a corporate issue forms a part of what some researchers see as the emergence of a new green techno-economic paradigm (Kemp, 1994; Freeman, 1992). The greening process is seen as revolutionary in character, entailing a paradigm shift resulting in major cognitive and technological changes. This paradigm change is expected to have thorough structural effects for the whole economy. But how does this green trend, pushed forward by the political system, diffuse on the market and affect interfirm dynamics?

This chapter investigates how firm interdependencies and the related interfirm learning influence the greening process. Empirical surveys state that there are major discrepancies in firms' strategic response to greening (Madsen and Ulhøi, 1995, 1996; Malaman, 1996). So far, only a minor though rising extent of firms are developing proactive environmental strategies. Firms are affected very differently by environmental issues, depending partly on structural factors such as the nature of their technologies (some sectors are inherently more polluting than others) and the size of the firm, but also on their history, noticeable the environmental legislation or NGO pressures they have been exposed to. In the current transition phase,

T.J.N.M. de Bruijn and A. Tukker (eds.), Partnership and Leadership, 103–119.

where the new green paradigm is only beginning to become integrated in the economy, the greening process is extra asymmetric. There is thus an unequal distribution of green incentives, green strategies and green capabilities among firms on the market. How does that affect the interfirm greening dynamics?

This chapter seeks to forward an economic organisation perspective on the greening process. Such an approach sheds light on how the greening of the economy is affected by the difficulties in organising the necessary co-ordination and learning processes within and notably between (asymmetrically greening) firms on the market. It is suggested here, that *system failures* on the market are important factors shaping the greening process. Interfirm learning processes hence gain significance in efforts to understand retardation in industrial greening processes and ways of overcoming it.

The chapter presents an analytical framework of interfirm learning drawing on innovation oriented neo-institutional economic theories. The green trend may be analysed similar to other paradigm and technological trajectory changes on the market, such as the introduction of electricity, new business practices related to the IT-technology etc. which innovation theory seeks to address already. But investigating the greening of industry does have its particular characteristics. The greening of industry is an innovation process that is distinct in character in four ways: it is radical, highly systemic, associated with considerable information problems and very politically influenced (Andersen, 1999). The greening process therefore entails a mixture of lower order learning (exploitation) and higher order learning (exploration) and incremental as well as very radical innovation. Consequently, the greening of the economy gives rise to high interfirm co-ordination needs.

The empirical question is then, if this asymmetry in green firm response *and* the highly systemic and radical nature of the green innovation process mean that firms do make each other go green as a part of the economic process. Do the more green firms seeking to attract green rents make the less green firms become greener? Are firms held back because some conservative firms or firms lacking green capabilities function as bottlenecks? Or are sectors, production chains, regions or other collections of firms swept forward together in a common collaborative green learning process, entailing a consolidation and assimilation of their (greener) strategies? A key theoretical question related to this discussion is whether the interfirm greening process primarily is a collaborative or conflictual process. In taking an economic organisation perspective the chapter seeks to encompass both these aspects of the interfirm green learning dynamics.

The chapter outlines the contours of a knowledge based theory of (green) interfirm learning. The framework should inform us on when we have interfirm learning and why it differs in intensity in different firm relations. In this short chapter, there is, however, only room for some main elements of the framework and emphasis is here placed on the incentive side while the cost side discussing what factors limit interfirm learning is neglected.

The chapter first briefly discusses interfirm learning from the perspective of economic organisation. Thereafter, the three suggested processes of interfirm learning are presented, respectively capability accessing, ex ante co-ordination and ex post co-ordination. The discussion is illustrated by some empirical examples of green interfirm learning between the Danish paper mills Danisco paper and Stora Dalum and their suppliers, customers and competitors[1].

2. A THEORETICAL APPROACH TO INTERFIRM GREENING

When it comes to discussing interfirm greening, there is much confusion as to what kind of processes are going on and as to the direction(s) of change. For instance what do we mean with partnerships or collaboration, which forms do they take on, when do they occur and between whom? There is clearly a need for building thorough conceptual frameworks, which may form the basis for coherent empirical work.

Much work on green interfirm learning takes a very simplistic perspective on the interfirm dynamics, e.g. discussing only formal collaboration or environmental requirements between firms. Current research on interfirm greening mechanisms is limited and largely managerial and prescriptive in nature. The focus is on the potential for establishing green integrated chain management; that is, how firms best communicate green issues to other firms and apply green pressures or demands to other firms (e.g. Cramer, 1994; Green and Morton, 1996; Heiskanen, 1997). Interfirm innovation processes are little touched upon. Other more evolutionary researchers discuss green technological co-operation between firms related to the development of specific innovations (Den Hond and Groenewegen, 1993; Cramer and Schot, 1993). Cramer and Schot (1993)

[1]The paper is based on the authors Ph.D. dissertation (Andersen, 1999). The dissertation is primarily theoretical, but entails also a thorough empirical study based on 27 interviews in 6 firms within the paper cluster. The Danish papermill Stora Dalum referred to in the text has since the investigation changed owner and is now called Dalum Papirfabrik. The dissertation may be purchased at Samfundslitteratur, Rosenørnsalle 9-11, 1970 Frederiksberg, Denmark.

thus point to the need for 'environmental co-makership' between firms to achieve green technological innovations. But from these studies we know little of the drivers for the interfirm greening processes, nor of the complexity of the interfirm greening dynamics.

This chapter links interfirm greening mechanisms to theories of economic organisation in an attempt to identify the different kinds of interfirm learning processes going on as the market is turning green. According to this approach the competitive process is inherently a mixture of interfirm collaboration and conflict (Penrose, 1959; Loasby, 1996, 1999). As firms and markets change firms need to constantly reorganise their production and learning, entailing at times interfirm collaboration and at times conflicts. Both are essential elements of learning processes.

The chapter takes a constructive approach. It focuses on theory development and in presenting the core elements of a conceptual framework of interfirm (green) learning rather than in criticising existing theoretical or empirical work. Basically, interfirm learning is seen here as taking place through both adaptation and conscious capability building processes, as will be elaborated on later. The theoretical development is inspired by an empirical analysis based on in-depth case studies of green interfirm learning processes in the paper chain.

The theoretical perspective pursued here may be designated the 'knowledge based theory of the firm'. In accordance with the newer tendencies within the innovation oriented economic theories, the chapter seeks to synthesise the strategy approach of the resource based theory with 'evolutionary economic theory' in order to intersect notions of economic organisation, learning, strategy and dynamic market processes.

In doing so, more emphasis is placed at the firm level than usually within evolutionary economic theory. Evolutionary economic theory has a strong focus on explaining the broader phenomenon of paradigm changes and lock-in (Dosi, 1982, 1988 et al.), but neglects explaining firm strategic behaviour. While resource based theory, or the theory of the firm, usually concentrates on explaining firm specialisation and integration, it is argued here that it can equally well provide a theory of interfirm learning. More emphasis is placed here on collaborative forms of economic organisation than is usual within resource based theory. The core idea pursued in this chapter is thus to extend the knowledge based theory of the firm to the interfirm level. The knowledge base theory is used here to explain how larger groupings of firms, e.g. clusters, value chains, and regional and national innovation systems, react to change, such as the green trend.

More specifically the theoretical discussion contributes to building a more dynamic transaction cost theory. The approach joins with authors who are seeking to integrate transaction cost and capabilities explanations in the attempt at developing a more dynamic neo-institutional economic theory (Langlois, 1992; Teece, 1986; Casson, 1997; Silver, 1984; Loasby, 1999). Firms and markets are seen as alternative systems of organising production and learning. Finally, the chapter also contributes to building a stronger microtheoretical foundation for the (national) innovation systems approach, (e.g. Lundvall, 1992).

3. THE ECONOMIC ORGANISATION OF PRODUCTION AND LEARNING

Many studies of both informal and formal interfirm learning, also those outside innovation theory traditions, implicitly or explicitly apply a capability explanation of interfirm learning. Thus it is commonly assumed that, at base, firms do interfirm learning because they want to access other firms' capabilities (possibly designated skills, know-how or knowledge). But we need an explanation for why firms choose to access other firms' capabilities rather than building these themselves or alternatively integrate with those firms who hold the desired capabilities. Interfirm learning must be seen within the overall organisation of production and learning on the market.

In the face of change markets may be contemplated of as a system for the co-ordination of the growth of knowledge (Loasby, 1996). The firm is a domain for learning within which resources are developed and deployed (Penrose, 1959). Firms from this perspective may be seen as a function of the degree of specialised knowledge brought under control and the accessibility to knowledge outside the firm, as explained notably by Richardson (1960, 1972, 1975). Richardson, in his seminal article "The Organisation of Industry" (1972), made a distinction into two categories of productive activities. Activities requiring the same or closely related capabilities are 'similar', while 'complementary', i.e. interdependent activities are typically situated in adjacent stages of production. This argument provides a knowledge based theory of the firm. There is a tendency for firm activities to centre around specialised knowledge or 'similar capabilities' which increases the co-ordination need on the market (Richardson, 1972). Firms only have the capacity to carry out similar activities based on similar capabilities, since incorporating 'dissimilar' activities would lead to diseconomies of scope and/or increased information costs. Firms will often benefit from spinning-off dissimilar activities leaving these to other firms (Richardson, 1975). Thus there are limits to the capabilities a firm can handle efficiently; an argument that forms the basis of modern resource based theory.

But Richardson's argument can also be extended to inform us on the relative distribution and value of capabilities between firms and thus on the incentives to do interfirm learning from a capabilities perspective. There are, presuming human rationality is bounded, limits to how much knowledge should be brought under firm control. Accessing them on the market may be much more attractive than control. They may sometimes be more effective capabilities if the firm does not try and control them, but leaves decision to those who know better and instead engages in some form of collaboration on the market. Access, rather than control, is particularly attractive when it is uncertain what capabilities will prove valuable in the future, for a firm can access much more than it can control (Loasby, 1996).

In Table 1 Richardson's argument is extended to discuss which activities a firm should co-ordinate internally and which to subject to co-operative efforts and with whom.

Table 1. Linking the division of labour to the division of capabilities

Activities	similar	dissimilar
complementary	internal co-ordination	vertical interfirm learning
independent[2]	horizontal interfirm learning	no interfirm learning

Source: Andersen, 1999

The capability concept is implicit in the table, as it, in Richardson's definition, is entailed in the 'similar/dissimilar activities'. The table should be interpreted as a dynamic model, which outlines some very basic incentives for firms to do interfirm learning, respectively integration in different firm constellations. The degree to which this actually happens depends also on other factors such as the costs of doing interfirm learning. Below in the discussion of the three suggested processes of interfirm learning the table will be explained more closely.

[2]The parameter 'independent activities' illustrates that between some firms the production needs not to be directly co-ordinated, e.g. between firms within the same sector (when the activities and capabilities are also similar) or between firms in different sectors not engaged in a division of labour with each other (when activities and capabilities are also dissimilar). It should be noted that the term complementary/independent activities refers strictly to the division of labour, where as there are coordination needs on entrepreneurial and investment plans also between firms with similar activities (Richardson, 1960) - in the latter sense the activities are not independent.

4. THE THREE PROCESSES OF INTERFIRM LEARNING

This section suggests three different interfirm learning processes based on empirical investigations of green interfirm learning in the Danish paper cluster (Andersen, 1999). Interfirm learning I see as made up of the direct - or interactive - information flows occurring as a part of either capability building or adaptation processes between firms. I use the term *capability accessing* to refer to interfirm intentional search aimed at capability building, that is capability building through knowledge exchange or collaborative knowledge generation with other firms. *Learning through adaptation* I see as the information exchange, in the form of communication, persuasion and teaching associated with the co-ordination of productive activities. Furthermore, I am proposing that there are two different types of co-ordination processes leading to interfirm adaptation, respectively *ex ante co-ordination* and *ex post co-ordination.* The latter distinction serves to emphasise the time element of when in the innovation process firms co-ordinate their innovative activities (Andersen, 1999).

The three interfirm learning processes are closely intertwined. Central for the distinction between the interfirm learning processes are the firms' immediate incentives to engage in interfirm learning. The point made here is that all the three interfirm learning processes contribute to the overall interfirm learning, including interfirm greening.

4.1 Capability accessing

Capability accessing refers to the linking up between given firms with the purpose of building capabilities; it is 'co-operative search', entailing knowledge exchange or actual shared knowledge building (collaboration). It may take on formal partnerships and collaboration, but informal knowledge exchange is much more widespread than formal partnerships (Allen, 1977; Von Hippel, 1987; Dodgson, 1993; Conway, 1995; Coombs, 1996).

I propose that there are three kinds of interfirm learning constellations. Firms have incentives to access:

a) The capabilities of firms with *similar activities* because their capabilities are also similar (e.g. paper industry- paper industry);
b) The capabilities of firms whose *activities and capabilities are complementary* (e.g. paper industry - chemical industry); and
c) Those whose activities are independent but where the *capabilities are (nearly) similar* (e.g. paper industry - other process industry).

Between other firm constellations there are no incentives to do capability accessing. The three firm constellations differ with respect to the overlap between adaptation and capability access needs and the need for protection of the intellectual property. The *green capability accessing process* represents the knowledge migration between firms horizontally and vertically, related to their green technical but also organisational (green management systems) capability building.

Examples of interfirm green capability accessing
The first example concerns the Danish paper mill Danisco Paper[3] which produces recycled corrugated paper. Competition is hard within the paper industry, and many Danish producers have been forced to close down during the 1970s and 1980s as they could not compete with the much larger paper producers in e.g. the neighbouring Sweden, Finland and Germany. Danisco Paper competes on conventional productivity, that is making their paper machines run faster than their competitors. The firm is very innovative, having some of the fastest paper machines in the world, but it has very little actual R&D. Although the mill belongs to a large Danish group, they are pursuing a learning strategy of maximising the external knowledge accessibility. They rely heavily on formal and informal knowledge exchange with other firms for their capability building, notably their suppliers, and it is crucial for their competitiveness to be selected as learning partners by other competent firms. For Danisco Paper knowledge sharing within the group is limited, partly because of varied capabilities and partly because it is not encouraged.

When experiencing problems with their water emissions limit in the mid 1990s, the firm took the drastic decision to close the water system completely, something only a few firms in the world had done at the time. In this radical technical innovation process they were aided by important advice from their chemical and equipment suppliers who had experiences from some of the other pioneering firms within this technology. Furthermore, at the early stage Danisco Paper visited and engaged in informal but very open knowledge exchange with a couple of these other pioneering firms (their European competitors). Danisco Paper then developed their version of the technology in close collaboration with their two main chemical suppliers, as many chemicals and the usage of the chemicals were changed but also for instance the water usage on the paper machine had to be redefined. This collaborative innovation process still continuous as the very difficult technology is being refined.

The chemical suppliers state that participating in this co-operative green technology development is equally important to them, as they need to build

[3]Danisco Paper is a part of the large Danish food and packaging group Danisco.

up their capabilities in the field and see Danisco Paper as a very competent learning partner. They expect all paper mills to want to close their water circuit shortly. They see a rapidly growing market for closed water systems in the process industries, and more generally for green products. In this environmentally very sensitive industry possessing green capabilities is seen as essential for their selection as transaction and learning partners. Although the sector generally invests heavily in R&D, they are also very dependent on collaborative search with their competent customers.

Danisco Paper decided to undertake the water closure venture although green profiling is somewhat, but not very, important to the firm because they had the confidence that they fairly easily could obtain the needed capabilities to accomplish this task. Staying in the technological forefront and being attractive learning partners is essential to their competitiveness, and they see green technologies as forming a part of this (Andersen, 1999).

The second example concerns the Danish paper mill Stora Dalum[4] that developed in 1990 an innovative 100 per cent recycled high quality printing paper. The big Stora Group, one of the largest paper groups in the world, had recently after the initial recycled paper innovation bought up the mill. The Stora Group practices a learning strategy of considerable self-sufficiency and protection of the intellectual property rights. The firm has several central R&D departments, which are available to the different mills around the world, and encourages strongly collaboration and knowledge exchange among the sister mills. In Stora Dalum this has meant that their own development department has been reduced and that the mill is relying heavily on Stora knowledge sources. The firm experienced in the first years severe technical problems with the new recycled paper due to impurities in the paper. The Stora R&D departments were engaged for several years in improving the Stora Dalum recycled paper, which led to important improvements (see though also the story under '*Green ex post co-ordination*' on this).

Stora Dalum has also been working hard on closing the water system, and has succeeded in reducing the water consumption considerably. This innovation process, involving several innovations, has taken place noticeable in close collaboration with a German and Swedish sister mill with similar problems, with the German mill as the initial innovator. Collaboration with the main research centre of Stora, which has several research projects on

[4]Stora Dalum was at the time of the investigation part of the Swedish Stora Group, one of the largest paper groups in the world. The mill is now the independent Dalum Paperfabrik. However, here the name at the time of the investigation is used, since the affiliation to the Stora Group is a key issue.

water saving, has also been important but not as important as the knowledge exchange with the sister mills.

Stora Dalum's chemical suppliers have also been involved in the innovations though to a lesser extent. Stora Dalum makes use of their pilot plants for testing their innovations, since, as they say, they pay for the service indirectly so they might as well use it, just as all their competitors do. Also the suppliers are allowed access to tests in Stora Dalum's laboratories. The equipment suppliers have not been decisive, but have come up with some useful ideas and know how as they tell how it is done at other mills (Andersen, 1999). Stora Dalum has only gained important knowledge from one competitor on the water issue, and that was from a visit at Danisco Paper and a discussion on their water closure.

In relation to developing their environmental management system, Stora Dalum has participated in a lifecycle analysis project of their core customers, the Danish printing industry. During that process the participating firms have been exchanging know how on environmental conditions in respectively the paper and printing industries and Stora Dalum has gained insights on how to carry out lifecycle analysis.

4.2 Ex ante versus ex post co-ordination

In distinguishing between ex ante and ex post co-ordination the question raised here is when firms co-ordinate their innovative activities in the innovation marathon? Does it take place early on, influencing the creation of variety, (ex ante co-ordination), or does it take place later on, seeking to make the varieties (innovations in different firms) compatible (ex post co-ordination). The distinction thus serves to emphasise the thus for neglected time element in the firm-market interaction.

Ex ante and ex post co-ordination refers to two different strategies. *Ex ante co-ordination* refers to the attempts to secure co-ordinated adaptation between firms with complementary activities. The purpose is to secure that market capabilities are always there when needed and that interrelated firms stay on the same wavelength on parameters critical for competitiveness (capabilities, image etc.). Ex ante co-ordination emphasises thus the ongoing but conscious adaptation attempts between producers and users, involving either co-operative or coercive communication of information on user demands and product properties. Ex ante co-ordination thus influences the very early stage of the innovation marathon.

Empirically, the *green* ex ante co-ordination consists mainly of the green demand setting processes, the explicit pressures, requests and inquiries, between suppliers and customers. Included in this process is the institutional response of firms to the requirements, the 'green profiling' which is undergoing change as firms invest in green information standards, such as certified management systems, eco-labels, environmental product declarations, environmental reporting, etc., as they seek to economise on their green transaction costs (Andersen, 1999).

Examples of interfirm ex ante co-ordination

When the Danish paper mill Stora Dalum in 1990 developed their recycled high quality printing paper it was not as a response to customer needs, but in an attempt to create a new niche market, as the small paper mill found it very difficult to compete on conventional productivity. The firm has experienced severe problems with marketing the recycled paper as the wholesalers (even those within the same corporate group) in the beginning had little insights and interest in the "greenness" of this paper. There were lacking information channels and codes particularly on green issues between Stora Dalum and their wholesalers. The environmental management systems had only to a limited degree entered into the sales and purchasing departments in the paper chain and persuasion and teaching between the transaction partners on environmental issues was therefore limited. There was a green marketing dilemma, also within the corporate group. The wholesalers were reluctant to promote the greenness of the recycled paper, as it would make the other paper products look less favourable.

Stora Dalum has invested extensively in green information standards and material, e.g. eco-labels, eco-datasheets, certified management systems and environmental reporting to improve their environmental profiling. Stora Dalum has experienced an increase in the green inquiries as they strengthened their green profile, and were met by rising demands on their environmental performance from their customers. In Stora Dalum expectations evolved that they must have an overall very high environmental performance in order to market their product as green, which includes rising demands on their suppliers' environmental performance.

Danisco Paper developed in 1956 a 100 per cent recycled corrugated paper product, but not as a green product but for cost reasons. Secondary fibre was cheaper than primary fibre. They still experience generally weak green demands from their customers, the corrugated board industry, but more from their customer's customers, the varied users of corrugated boxes. Danisco Paper has, however, become green more because of the rising environmental debate related to the recycling issue. As society came to consider their product as green, so did Danisco Paper. Today the firm has developed advanced environmental strategies and is for instance EMAS certified.

The corrugated board industry is an environmental low impact sector, which has been late in developing environmental strategies. Danisco Paper works continuously to teach their customers of the greenness of their products. Also here has the technical customer service overtaken much of the green complex and rising communication with customers. Lately, the corrugated board industry has agreed to participate in a voluntary agreement to enhance the recycling of corrugated board. Danisco Paper participated actively in this process, not with the individual customers but at the sector level participating in negotiations with the industrial associations. Danisco paper set a guarantee that they would buy all the waste paper they could collect, in order to support their customers in their green marketing. The corrugated board industry were initially very reluctant towards the agreement and the whole idea, but has changed their attitude, and is now engaged actively in green marketing for the first time. The sector has produced 'an environmental response catalogue' to be used by the sales people.

Danisco Paper participated in developing an extensive LCA methodology on corrugated paper in co-operation with the other big European producers of corrugated paper and corrugated boxes in order to try and influence on the LCA standard. The LCA is seen as an important tool in the competition with the producers of plastic packaging, where environmental issues increasingly are a competitive factor. Danisco Paper is participating actively in the debate on the plastics vs. corrugated boxes controversy as it is seen as affecting the whole market for corrugated board (Andersen, 1999).

Ex post co-ordination focuses on co-ordination efforts arising from interfirm conflicts related to incompatible products or activities. The variety of innovations, i.e. occurring ex post the contemplation of a given product innovation, are sought made compatible. The conflicts arise when innovation renders the market capabilities obsolete and incompatible products or activities need to be co-ordinated. Ex post co-ordination takes place in the mature phase of the innovation marathon. It is carried out through the exercise of authority in the form of persuasion or teaching.

The *green* ex post co-ordination activities consists of persuasion and teaching on environmental issues between firms with complementary but incompatible products or activities. This co-ordination need may go beyond the actual value chain (Andersen, 1999).

Examples of green ex post co-ordination

After the initial development of their recycled paper Stora Dalum experienced problems with the quality of the paper product. The firm tried partly to solve the technical problems by engaging heavily in R&D, though not even the advanced R&D department of the great Stora group was able to solve all the problems. The firm also engaged in persuasion with other firms whose products are influencing in a negative way on the recyclability of their paper (and in this respect their products are complementary although they are not transacting partners), the producers of adhesives and envelopes, in order to solve some severe technical problems[5], and has also sought to engage in teaching (capability building) with these by applying their know-how in joint research projects to overcome the problem. The results were limited though due to diverging interests. The other firms saw little market potential in the recycled paper for them.

Stora Dalum also engaged similarly in persuasion and teaching with a large user of their paper, Xerox, one of the largest copier producers world-wide and a firm with advanced environmental strategies. Serious technology development and collaboration was undertaken, but technical incompatibilities remain between the recycled copier paper and the Xerox copier machines.

Xerox practices a strategy of being a major paper producer without paper machines (minimises self-sufficiency). Xerox develops new paper products and makes a range of paper producers produce the paper for them on contract. This allows Xerox to obtain maximum innovative capacity and continuously move on to new products. Xerox thus had several different recycled papers under surveillance and one on contract (i.e. produced with the Xerox name) at the time of the co-ordination efforts with Stora Dalum. Failing to get the Xerox approval in this early stage of market entrance has severely limited the sales opportunities and the reputation of Stora Dalum's copier paper, and has made Stora Dalum shift their interest into other kinds of printing paper, with greater success.

These examples illustrate the very complex and far-reaching technical interdependencies often related to green innovations, and the great problems of overcoming them (Andersen, 1999).

4.3 The relationship between ex ante, ex post co-ordination and capability accessing

The argument made here is that all the three interfirm learning processes contribute to the overall interfirm learning, as the empirical interfirm greening examples illustrate. The interfirm collaborative learning efforts, the

[5]The problems are mainly due to 'stickies', which are adhesive spots occurring in the recycled paper, which originate primarily from the adhesives on envelope waste paper. The stickies disrupt the paper production process, but also the quality of the paper because of spots reducing the readability or because the spots disturb the copying process.

information exchange and pressures resulting from these three interfirm learning activities influence the rate and direction of innovation by shaping the firms' knowledge bases, their heuristics and their entrepreneurial expectations.

This shaping may take place at many levels of the firm. Much of the interfirm interaction takes place at the shop floor entering imperceptibly into firm routines and heuristics and may take a while before it reaches the overall strategic level of the firm.

The distinction of the three interfirm learning processes sheds light on the complex dynamic firm-market relation. Firms have different strategic options in organising their production and learning on a changing (greening) market. A neglected issue is that the co-ordination and capability building processes to some extent are alternatives. A firm may choose between building certain capabilities themselves, or engaging in ex ante or ex post co-ordination in order to get other firms to undertake a part of the innovation. Firms may also choose between pursuing either a primarily ex ante or ex post co-ordination strategy towards changes in their selection environment.

The empirical findings of green interfirm learning in the paper cluster indicate that both ex ante and ex post co-ordination measures are applied in the co-ordination of the complementary activities, and thus that adaptation at times takes place in a very co-operative manner, at other times more adversarial. The firms portrayed all apply a mixture of ex ante and ex post co-ordination strategies. Neither of these strategies are pursued very systematically. The firms have insufficient knowledge, organisational set-up and strategies for dealing with the changes in their selection environment related to the greening process.

5. CONCLUSIONS

This chapter seeks to link interfirm greening mechanisms to economic organisation. The chapter investigates how firm interdependencies and the related interfirm learning influence the greening process. The empirical study of the greening of industry represents an unusual innovation study of how new techno-economic paradigms, including new heuristics, strategies and technologies at the firm level, diffuse between firms. The empirical implications of the approach forwarded here is that the greening of the economy is affected by the difficulties in organising the necessary co-ordination, collaboration and overall learning processes within and notably between asymmetrically greening firms on the market. It is suggested here, that *system failures* on the market are important factors shaping the greening process. Interfirm learning processes hence gain significance in efforts to understand retardation in industrial greening processes and ways of overcoming it.

The chapter suggests that the knowledge-based theory of the firm provides a fruitful framework for analysing interfirm greening dynamics. A mixture of capabilities and contractual explanations has been provided in order to explain when firms undertake different kinds of green interfirm learning processes and with whom. It is a microtheoretical framework that outlines the incentives and costs structures that firm's experience as they seek to organise their production and learning efficiently in a greening economy.

According to the suggested framework interfirm greening results from both collaborative capability building and adaptation processes between firms. Furthermore, it is suggested to make a further distinction of the adaptation processes, respectively into ex ante and ex post co-ordination processes, hereby bringing attention to when in the innovation process firms co-ordinate their innovative activities.

The qualitative nature of ex ante and ex post co-ordination processes is very different. The former is co-operative, the latter adversarial. The framework forwarded here thus illustrates the mixed co-operative and adversarial nature of the economic process, and has brought attention both to the system failures and the strong co-ordination and collaboration needs related to the greening of industry. This discussion emphasises the key role of interfirm dynamics for the occurrence and diffusion of new market trends, such as the greening of the market.

The greening of industry, it is here suggested, may only be explained by incorporating all the three suggested processes of interfirm learning: capability accessing, ex ante co-ordination and ex post co-ordination. Together the three processes encompasses how firms seek to organise their production and learning efficiently across interdependent firms which face each their conditions and opportunities towards the greening selection environment. The approach forwarded here emphasises the interplay between the three interfirm learning processes and calls for the need to see these in their interplay and within the wider context of economic organisation.

REFERENCES

Allen, T.J (1977) *Managing the Flow of Technology*, Cambridge MA: MIT Press.

Andersen, M.M. (1999) *Trajectory Change through Interorganisational Learning – On the Economic Organisation of the Greening of Industry*, Copenhagen Business School, Ph.D. series no. 8, Copenhagen: Samfundslitteratur.

Casson, M. (1997) *Information and Organisation*, Oxford: Clarendon Press.

Conway, S. (1995) Informal Boundary-spanning Communication in the Innovation Process: An Empirical Analysis, *Technology Analysis & Strategic Management*, 7(3): 327-343.
Coombs, R. et al. eds. (1996) *Technological Collaboration - The Dynamics of Cooperation in Industrial Innovation*, Cheltenham: Edward Elgar.
Cramer, J. (1994) *Experiences with Integrated Chain Management in The Netherlands*, Working Paper, Apeldoorn: TNO Centre for Technology and Policy Studies.
Cramer, J. and J. Schot (1993) Environmental Comakership among Firms as a Cornerstone in the Striving for Sustainable Development, in *Environmental Strategies for Industry* K. Fischer and J.Schot eds., 311-328. Washington D.C.: Island Press.
Den Hond, F. and P. Groenewegen (1993) Solving the Automobile Shredder Waste Problem: Coorperation among Firms in the Automotive Industry, in *Environmental Strategies for Industry* K. Fischer and J.Schot eds., 343-367. Washington D.C.: Island Press.
Dodgson, M. (1993) *Technological Collaboration in Industry - Strategy, Policy and Internationalization in Innovation*, London: Routledge.
Dosi, G. (1982) Technological Paradigms and Technological Trajectories: A Suggested Interpretation of the Determinants and Directions of Technological Change, *Research Policy*, 11:147-162.
Dosi, G. et al. eds. (1988) *Technical Change and Economic Theory*, London: Pinter Publishers.
Freeman, C. (1992) *The Economics of Hope*, London: Pinter Publishers.
Green K. and B. Morton (1996) *Green Purchasing and Supply Policies: Do they Improve Companies' Environmental Performance?* Paper presented at the 5th International Conference of the Greening of Industry Network, November 1996, Heidelberg.
Heiskanen, E. (1997) *Environmental Product Improvement: Key Actors and Information Flows in the Product Chain*, Paper presented at the Nordic Business Environmental Management Network Conference June 5-7 1997, Tuohilampi.
Kemp, R. (1994) Technology and the Transition to Environmental Sustainability, *Futures*, 26(10): 1023-1046.
Langlois, R.N. (1992) Transaction Cost Economics in Real Time, *Industrial and Corporate Change*, 1:99-127.
Loasby, B. (1996) The Organisation of Industry, in *Towards a Competence Theory of the Firm* J. N. Foss and C. Knudsen eds. 38-53. London: Routledge.
Loasby, B. (1999) Understanding Markets, in *Knowledge, Institutions and Evolution in Economics* 107-129. London: Routledge.
Lundvall, B. ed. (1992) *National Systems of Innovation*, London: Pinters Publishers.
Madsen, H. and J. Ulhøi (1995) *Sustainable Corporate Management in Denmark*, DEMS Working Paper 3, Institut for Informationsbehandling, Århus: Århus Business School.
Madsen, H. and J. Ulhøi (1996) Environmental Management in Danish Manufacturing Companies: Attitudes and Actions, *Business Strategy and the Environment*, 5:22-29.
Malaman, R. (1996) *Technological Innovation for Sustainable Development: Generation and Diffusion of Industrial Cleaner Technologies*, Milano: Fondazione Eni Enrico Mattei.
Penrose, E.G. (1959) *The theory of the Growth of the Firm*, New York: Wiley.
Richardson, G.B. (1960) *Information and Investment*, 2nd edition 1990, New York: Oxford University Press.
Richardson, G.B. (1972) The Organisation of Industry, *The economic Journal*, 82:883-896.
Richardson, G.B. (1975) Adam Smith on Competition and Increasing Returns, in *Essays on Adam Smith* A.S. Skinner and T. Wilson eds. Oxford: Clarendon Press.
Silver, M. (1984) *Enterprise and the Scope of the Firm. The Role of Vertical Integration*, Oxford.

Teece, D. (1986) Profiting from Technological Innovation: Implications for Integration, Collaboration, Licensing and Public Policy, *Research Policy,* 15:27-44.
Von Hippel, E. (1987) Cooperation between Rivals: Informal Know-how Trading, *Research Policy*, 16:291-302.

Chapter 7

CONDITIONS FOR LEARNING PARTNERSHIPS

The Introduction of Organic Food by a Swiss Retailer

SIMONE MAIER

1. INTRODUCTION

Many scholars work on the topic of collaborations because they are seen as a necessary and appropriate means to foster a sustainable development (see Hartman et al. 1999 and in this volume, for an overview). Collaboration is usually assumed to enhance learning processes between the partners towards a sustainable development (Heller, 1998; Clarke and Roome, 1999, in this volume). But as this case of collaboration along the food chain reveals, several conditions regarding the choice of partners, the division of power among the partners, as well as the interests at stake have to be fulfilled in order to make a collaboration a successful learning partnership.

Nutrition is an important domain regarding sustainable development in various respects and as such referred to in several chapters of Agenda 21 (UNCED, 1993). To introduce food from organic farming to the food chain is one method to support sustainable development. Although organic farming is not unchallenged in its ecological effects, which is due to neglectful or inappropriate research design of tests (Jungbluth, 1999; FAO, 2000), the FAO (2000) admits that regarding the use of chemical pesticides and fertilisers as well as effects on biodiversity organic farming is likely to have advantages compared to conventional production methods. It thus is

T.J.N.M. de Bruijn and A. Tukker (eds.), Partnership and Leadership, 121–134.

also likely to provide more healthy food as it contains less residues from chemical treatment. Regardless of the doubts on the details, organic farming is already being subsidised in several countries in order to make more farmers convert to this production method. In Switzerland, subsidies for organic farming were introduced in 1993 as one part of a successive fundamental reformation of the agricultural policy which will be fully implemented in 2002. But apart from the political support, organic food products also need a market.

Unfortunately, the organic production method does not necessarily show at sight of the product, especially not with processed food. Therefore, a label is needed to confirm the quality of the product and create credibility at the customers' side. Labelling is best done by third parties in order to resolve all doubts about potential interests involved in the quality claim (Welford, 1995:167). Labelling organisations for organic food traditionally are situated on the organic farmers' level because historically, the labelling usually started with establishing shared standards for organic farming. Then, the farmers realised that it would be useful to include the further steps of the product chain: food processing and trade too in order to secure the quality standards that have been set on the farming level all through to the customer. But this also meant to acquire new expertise and interact with new partners, as will be shown in the case analysed here.

The Swiss labelling organisation 'Bio Suisse' also is the umbrella association of the Swiss organic farmers. Bio Suisse owns the organic label 'Die Knospe' (the Bud). Before the labelling of organic products was regulated by a Swiss federal regulation[1] the Bud was the only widespread quality standard for organic food in Switzerland. This has led to a quasi-monopoly for the Bud and Bio Suisse on the Swiss food market. Another development yet enforced the monopolistic situation: Coop, the second-largest Swiss retail trading company introduced in 1993 its own retail brand for organic products, the Coop NaturaPlan, and decided to sell only organic products with the Bud under this brand name[2]. Coop thus acts as an important 'pull'- and 'push'-power on the Swiss food market: in order to develop the product range of its NaturaPlan brand, it asks food processing companies to produce organic food products. As Coop also runs its own restaurants, it also asks for convenience catering products.

One has, however, to admit that the total market share of organic products in Swiss retail trade has just reached two percent in 1999. In catering, it is still much lower. There are about ten large players in the else

[1]The Swiss federal regulation for herbal and vegetable organic products is established since 01/1998, for animal products since 01/2001.

[2]For a more detailed account on the development of the Swiss market for organic food and the labelling organisations, see Villiger et al. 2000: 96-108.

very fragmented Swiss catering market and only one of them apart from Coop follows something like a moderate push strategy for organic meal components. Both are licensees of the Bud label. Most single restaurants do not offer any organic meal components and the very few ones that do so are more oriented towards fresh production without much use of convenience products.

This is the background of the case which will be described in this chapter: The case company, Luxfood Productions[3], was asked to produce organic potato gratin with the Bud label for the Coop Restaurants. We will analyse the collaboration between Luxfood Productions and Bio Suisse in the labelling process regarding three aspects: the power constellation in the collaboration, the learning processes in the collaboration, and the question of missing partners in order to reveal conditions critical to the learning process.

The results presented here are taken from a qualitative case study (Yin, 1994) that was part of a research project performed between summer 1997 and spring 1999 in Switzerland[4]. We studied the introduction of organic products to 'conventional' organisations, i.e. such organisations that had so far processed only conventional food. The case referred to in this chapter was particularly focussed on the interaction with the labelling organisation. The case was chosen because it was explicitly not a success story but several difficulties had come up during the introduction of the organic product and in the collaboration with the labelling organisation.

Our interview partners were chosen by a 'snow ball system': we started with the person in the labelling organisation who was responsible for the labelling process. He was asked to provide a contact to food processing organisations where difficulties had come up during the introduction and name his most important interaction partner there. After the participation in the case study was agreed upon, we held the first interview with that person and asked our interviewee to identify other colleagues who were also concerned with the introduction of the organic products.

We did semi-structured interviews starting with a narrative part and introducing theoretical concepts only in the second half of the interview. The interviews were recorded and transcribed. A résumé was written from the transcriptions and sent to the interviewees in order to confirm or correct, where necessary, relevant data and a first rough interpretation of their statements. The transcriptions were analysed with the qualitative data

[3]This is not the real name, the company was granted anonymity.

[4]The project was called 'Organisational and inter-organisational learning processes towards sustainability'. See Maier et al. 1998 for an outline of the project plan and Maier 2001 for final results. It was one partial project of an integrated transdisciplinary research project entitled 'Sustainable Switzerland in the international context. Development of visions, strategies, and instruments for the need-field of nutrition.' See Minsch, Mogalle 2000 for the final report of the integrated project.

analysis software 'ATLAS/ti' (Miles, Weitzman, 1994)[5]. Furthermore, the results of the analysis were discussed with participants from both organisations in a workshop[6].

2. THE INTRODUCTION OF ORGANIC PRODUCTS AT LUXFOOD

Luxfood Productions is the smallest subsidiary of an affiliated group with its main business in the catering and hotel business. Luxfood Productions was originally founded in order to provide convenience products for the Luxfood catering branch. The subsidiary produces convenience products, mainly for catering. Some successful products are sold via retail too. Nowadays, the Luxfood restaurants don't have to buy all products from Luxfood Productions any more, the subsidiary competes with other catering convenience product manufacturers. The product range, of which ninety percent is deep frozen, contains among others various gratins, rolls, croissants, but also salad dressings, etc. The product policy concentrates on constant and high quality. The restaurants using Luxfood products have decided for convenience products but value good quality nonetheless. Prepared and served the products cannot be identified as deep-frozen convenience at first sight. The Luxfood restaurants' product philosophy from an outside perspective on their menu cards could be summed up as serving delicious and exotic food from all over the world adjusted to the Swiss taste and at reasonable prices. Organic products do not play any role in the group's strategy.

In fall 1995, Coop Restaurants inquires whether Luxfood Productions could provide a Bud-labelled organic version of the deep-frozen potato gratin that Coop already buys. The product manager agrees as he interprets this demand as a chance to gain some experience with organic products and also as a measure to ensure the business with this important client. As this is Luxfood's first organic product contact is established with Bio Suisse, in order to find out about the conditions that are demanded to obtain a license for the Bud label for this product. He is informed that ninety-five percent of the ingredients have to come from organic production and of Swiss origin. Also, there are some rules regarding the processing methods and packaging.

[5]The program allows to highlight quotations in e.g. interview texts, assign codes to them and run queries in order to filter the quotations linked to single codes or code combinations. It thus enables researchers to test hypotheses and build theory according to the 'grounded theory' method.

[6]For a more detailed account on the choice of the cases and on the research methodology, see Maier 2001.

The recipe of the product has to be submitted to the licensing committee which holds a meeting every three months. The product manager gets irritated over the long delay as Coop demands the product much earlier. After extended negotiations he arrives at getting a preliminary license that gets confirmed officially at the next licensing meeting. The organic potato gratin is delivered and Coop places several follow-up orders. This causes some problems for the purchase department as the market for organic potatoes is very tight. Most quantities are ordered before the cultivation period so that farmers and wholesale traders know how much potatoes they must provide. There is still overdemand for many organic products in Switzerland as not enough farmers have converted to organic production.

Later on, some few more clients can be convinced to order the organic potato gratin or other vegetable-pasta gratins that were developed in the meantime. However, a planned extension of Luxfood's organic product range beyond the gratins is difficult. The product manager suggests an organic 'Bircher Müesli' (a famous Swiss snack, a mix consisting of squished wheat, grated nuts, raisins, berries, and milk or yoghurt) to Coop and another large catering company. Both are interested and pleased at the suggestion to have it deep-frozen in order to maximise the convenience for their kitchen staff. The product manager has a first 'prototype' recipe developed and applies for the Bud license, which is refused by Bio Suisse. One of the Bio Suisse licensing rules is based on the assumption that the customer approaches an organic product with certain perceptions about its qualities and production methods. The rule says that this assumed customer perception must not be deceived. The reasons for the refusal given by Bio Suisse is that a customer assumes that a Bircher Müesli is produced fresh on the spot and not in a factory and then defrosted in the restaurant. The Luxfood staff is angered and frustrated at this incident.

After three years the whole turnover with organic products still remains low. Due to the high purchase prices of the raw materials and the higher production costs due to comparably small production lots and higher work intensity, the contribution margin remains on an unsatisfactory low level.

3. ANALYSIS

3.1 Power constellation

Bio Suisse has a very powerful position on the Swiss organic market, at least in the Coop retail channel which in 1999 amounted to about 43 percent of the whole retail market for organic food in Switzerland (Bio Suisse,

2000). Coop and Bio Suisse have formed a mutually re-enforcing partnership. Coop profits from the third party credibility that the Bud label assigns to the NaturaPlan product range. Bio Suisse profits from the enormous advertising budget that Coop spends in order to make the NaturaPlan products and thus also the Bud label known. The Swiss 'eating out'-market is quite fragmented with about ten large players with a market share of about 30 percent and many single restaurants sharing the rest of the market (Wüstenhagen, 1997). The two large companies that follow a moderate push strategy to offer organic products are Bud licensees (Bio Suisse, 2000b). The party coming between hammer and anvil are the suppliers, like Luxfood Productions. They are forced to take a license and abide by the Bud rules if they want to deliver organic products to those channels.

The Bud label is safe from possible reproofs of being a non-tariff trade barrier, as it is a privately owned label, not a state assigned one. And usually, one would argue that the owners of a private label could set whatever rules they like, because the applicants for a license need not take the contract as the label is not mandatory. However, the monopoly in the Coop channel does give rise to questions. It can be said in favour of Bio Suisse, that it is a very democracy-oriented organisation that tries to include all its stakeholders into important decisions, e.g. it assigns a right of speech to its processing licensees whereas no voting rights (which are only assigned to its farming members) in its general assembly. Also, it must be said that Bio Suisse so far has done much in order to ensure a flawless organic quality, with strict controls all along the product chain. Bio Suisse thus has played an important role in the quality protection and the marketing of organic products which so far has helped to make the Swiss market for organic products one of the best developed all over Europe[7]. However, the Bud label is not a perfectly third-party-assigned label as Bio Suisse is the Swiss organic farmers' association in the first place. Therefore the interest of the Swiss organic farmers remains an influence factor in the licensing process. Also, a certain scepticism towards industrial processing organisations and the application of industrial processing methods could be found: another criterion for the Bud label is the prevention of 'unnecessary' processing which is justified by the preservation of food quality. But the more the Bud label is demanded for convenience products, the more difficult it will become to decide whether a processing step is necessary or not. Here is a large room for interpretation which poses a danger for arbitrary decisions if not handled with care.

One must, however, not forget that setting quality standards is not only a normative question but also a marketing decision. The basic rules for organic

[7]Market share in Switzerland is at about two percent, in Germany at about one percent. The average market share in the EU is at about 0.5 percent. See Bio Suisse 2000a.

products have been fixed in a Swiss federal regulation and the criteria for the Bud label go beyond the rules of the federal regulation in several ways. This 'extra'-quality associated with the Bud label is honoured by Coop and by those Swiss customers who buy the Bud labelled products. Thus, Bio Suisse's power position in the Coop channel demands a clear decision by potential Bud label applicants in favour or against this channel. This might sometimes mean to abstain from some amount of turnover. But one also has to calculate the costs of the implied restrictions, like buying raw products from Bud label origin as well as some extra limits on food additives and processing methods. Also, selling Bud-labelled products to non-licensees means the loss of surplus costs as they are not allowed to use the label as a quality sign towards the customer. This is particularly important in the Swiss catering business with only two large catering organisations and about ten single restaurants being Bud licensees (Bio Suisse, 2000b). So, in the case of Luxfood Productions, it is rather doubtful whether the decision in favour of the Bud labelled products was a good one as they subordinated their product development to the 'Bud rules' for entering a very tight market.

3.2 Learning processes

Many overlapping or competing concepts of organisational learning processes have been developed so far (Argyris and Schön, 1978 and 1996; Hedberg, 1981; Probst and Büchel, 1994; Nonaka and Takeuchi, 1995; also Fiol and Lyles, 1985, for an overview). A definition was chosen for the context of this analysis that does by no means claim to be the only possible one. Rather it was rather developed regarding its contributions to the analysis of the empirical problems: in organisational learning processes, the knowledge base of the organisation is enlarged and/or changed, the competence of its members for solving problems and acting as organisational agents is improved, and the mutual mental framework of the organisational members is changed in order to maintain or improve the organisation's ability to succeed in its relevant context(s).

The collaboration between Bio Suisse and Luxfood Productions did provide different learning opportunities in the sense of the definition above. Taking the introduction of the Swiss federal regulation for organic products in 1998 into account, Bio Suisse was forced to improve its service quality in order to keep its licensees from switching to the 'federal organic definition' as a quality proof easier to obtain. The collaboration with Luxfood Productions did certainly stimulate the understanding of the Bio Suisse staff for the need of a fastened licensing process in order to shorten the applicants' 'time to market'. Apart from this learning opportunity on the process level, the collaboration with Luxfood Productions also gave the Bio

Suisse staff another insight into the philosophy of a 'conventional' industrial food processing organisation. The general trend to more convenience products that also applies to the organic market forces Bio Suisse more and more to collaborate with food processing organisations that have no experiences with organic products and therefore are not familiar with an organic product philosophy. Whether or not this particular collaboration with Luxfood Productions enhanced the acceptance of such organisations at Bio Suisse, it certainly gave the staff more experience with such 'newcomers', regarding the needs for information about suppliers of organic raw material in quantities suitable for an industry organisation, about processing methods both acceptable to Bio Suisse and suitable for industrial production, as well as for the necessity of a careful acclimatisation of such a 'foreign' organisation to the organic philosophy and the reasoning behind the Bud label criteria.

This last issue leads us to the lessons (not?) learnt by the Luxfood Productions staff. The main problem of the introduction of the organic products certainly was the question whether these would fit into Luxfood's product policy. As already mentioned, Luxfood in general claims a high quality standard for its products. In the collaboration with Bio Suisse, both had to learn that the own idea of high quality does not necessarily equal high quality in another organisation's standard. Whereas Luxfood Productions' idea of high quality could be called 'outward' quality, Bio Suisse rather follows an idea of 'inner' quality. The Luxfood 'outward' quality is associated with a flawless outward appearance, ever constant characteristics in the defrosting and heating process in the restaurant and always the same flavour. Undoubtedly, those characteristics cannot be reached without the help of several additives, such as flavour enhancers, stabilisers, and food colours. What is helpful to the one party however is forbidden by the other's standards. The Bio Suisse 'inner' quality is linked with the abolishment of chemical and artificial additives on the farming, as well as on the processing level. Accordingly, the high quality of a product in this case cannot as much be derived from the outward appearance as from tests proving significantly lower chemical residue (FAO, 2000), the higher content of natural nutrients, or the stronger 'vital forces'[8] of its raw materials. To conclude, both organisations set their priorities on different, but not necessarily mutually exclusive aspects of quality in order to establish customer loyalty. However,

[8]The juice of a fruit or vegetable is mixed with caustic soda and made mount through a wick in the middle of a filter paper that was prepared with silver nitrate. The juice reacts with the silver nitrate and builds a structured image. The juices from organic products build a more balanced structure than those of conventional farming. This is taken as an indicator for their superior 'vital forces'. (Balzers-Graf, 1991)

neither did get involved enough with the other one's perspective to discover the mutually supportive aspects of the two quality standards.

As Luxfood Productions is only the smallest subsidiary of the affiliated group and the turnover with the organic products remained a very low share of the overall turnover (less than 1 percent), the confrontation with a totally different quality standard did not lead to any changes on the organisational level but rather to personal irritation and problems. The staff involved in the launch of the organic products could not change the established quality criteria. This might only have been possible, had the market response to the organic product range been much bigger. They were rather forced to make the organic products produced according to the Bud label criteria fit as best into the established (and ISO 9000-certified) quality management system as possible. However, there were 'learning' processes on the individual level, as some collaborators did report that they had acquired a different view on the food products that they worked with every day. They said that they had learnt again that the products were food, "something that other people eat, and not screws" (Interview I).

It is a question of perspective, to decide what sort of learning processes occurred on the marketing level. Although it is proven by several consumer surveys both on the verbal and the action level (Arnold et al., 1999; Tanner et al., 1998; Braune and Franke, 1992; Matiaske, 1992), that price does play neither the only nor the decisive role for choosing a food product, neither in shops nor in restaurants, the catering companies themselves are very price-sensitive because there is a hard competition in the Swiss catering sector (Interview II). Therefore, the target group, mainly large catering companies that use convenience products, did not respond too enthusiastically. As a consequence the organic products were driven back to a very low scale marketing, only having the products on the sales lists without promoting them actively. In the Luxfood restaurants it was left to the individual restaurant manager to decide whether the organic Luxfood products should be listed in the menu or not and what kind of promotion was to be made. No attempts were made towards a joint strategy in order to reach better economies of scale and thus lower prices. So, one could argue that they haven't learnt to integrate the organic products successfully. But, seen from another perspective, the Luxfood marketers have learnt their lesson quite well: the consumers did not respond too enthusiastically to the organic products so far, the turnover remained low and the contribution margin was low too. The few clients willing to pay for the organic products are served with a low scale offer. The goal of making first experiences with organic products has been reached. Those can now be preserved on a low expenditure level until the market responds more friendly. It thus depends on the eye of the beholder to decide whether the processes within Luxfood Productions could be judged successful learning or not.

To conclude, we can see that the collaboration of Luxfood and Bio Suisse has not inspired any mutual learning processes in the sense of learning together and reaching beyond the level of departure of each partner. But each partner on its own has learnt some lessons nonetheless. Both partners have extended their knowledge base: Bio Suisse has gained more experiences regarding the needs of an industrial licensee, Luxfood Productions has acquired some knowledge about the Bio Suisse licensing rules and about how the supply market for organic food works. The action and problem solving abilities of the Bio Suisse members have been trained in the negotiation process regarding how to deal with a newcomer from the 'conventional' food processing industry. The Luxfood members have managed to fit the organic products into their existing product range as good as possible with as little effort as possible. This certainly provided some exercise in problem solving. The mutual mental framework of the Bio Suisse members may at least have been challenged regarding their critical approach towards industrial production and convenience products. The need for service orientation besides safeguarding the organic quality has certainly re-enforced the organisation's ongoing re-orientation towards a more marketing oriented approach. The members of Luxfood Productions seem not to have changed their mutual mental models at all as organisational agents. On a personal level some members changed their perception of their products as being 'food'. As organisational agents, the existing product strategy did frame the introduction of the organic products as not really fitting in with the conventional products. This mental framework was not challenged, the introduction of the organic products could rather be called an imposed strategy (Mintzberg and Waters, 1985) with some of the negative consequences for the available degree of freedom of decision-making. It was not tried to change any of the external structures in order to improve the organic products' market chances, which leads us to the last issue.

3.3 Partners missing – or wrong partners at all?

It seems that Bio Suisse was more of a necessary than a sufficient partner for Luxfood Productions in the introduction of its organic product range. Bio Suisse provided the label demanded by Coop. As we have already seen, the Bud label nowadays is an almost indispensable condition to enter the Swiss organic market, both in retail and in catering.

But on the marketing side, Luxfood Productions with this venture was heading almost straight against the direction of the 'house philosophy'. Most probably, the Luxfood product manager has underestimated the inhibiting effect of the conflicting product philosophies on the introduction process and the further development chances of the new products. With the marketing strategy of the group given, it was not possible to introduce the organic products to the Luxfood restaurants in a more powerful way thus using them as partners for a push strategy. He also has probably overestimated the market demand for the organic convenience products in relationship to the marketing efforts made. But this could not have been foreseen from the outset. In 1995, there was a certain euphoria for organic products, triggered by the reform of the agricultural policy as well as the entry of Coop into the organic market. Therefore, the experiences of Luxfood Productions should be considered as a necessary experience that could only unfold in the course of events. It is a characteristic of young markets, like the one for organic products, that the limits first have to be found.

But this does not necessarily mean that any organisation necessarily should have had the same experiences. The example of Coop shows that an organisation with a strategy that clearly builds its competitive advantage on organic products is able to create structures that make a market success possible. Certainly, this needs partners, as neither the supply nor the sales structures for organic products are already as developed and clearly shaped as for conventional products. And especially an organisation as Luxfood Productions, that is not in direct interaction with the consumers, needs partners that are, in order to push organic products. But – as we already have seen, Luxfood Productions did not have an own, genuine impetus to introduce organic products, it was merely asked by an important client. Consequently, both sales and contribution margin remained low as the active measures that would have been necessary for improving them, could not be taken against the group's general strategy. So, whereas Luxfood Productions on a business level was a good partner for Coop because it did supply the demanded product in time and in the demanded quality, it was not a good partner in the sense of a market development for sustainable catering products as it was bound by its group's strategy to refrain from activities in that direction.

4. CONCLUSIONS

Regarding the issue of power, the collaboration of Bio Suisse with its licensees certainly has favourable effects on the overall development of the Swiss market for organic products. The strict quality standards that are based among others on the customers' assumed expectations towards organic products are suitable to create credibility. The licensing condition to use mainly Swiss products contributes to the further development of the Swiss market for organic agricultural products. However, for a single food processing organisation, like Luxfood Productions, this private regulation can limit the diversification of the organic product range and thus the overall market development, as not every possible product is licensed.

Regarding the issue of learning processes, one finding certainly is that organisational learning needs organisational incentives and does not happen 'just by chance'. Bio Suisse had a clear strategy first regarding safeguarding the quality and developing the Swiss market for organic food which guided them through the decisions of the licensing process and second regarding the improvement of their service quality which made them open for speeding up the licensing process in order to shorten 'time to market' for their licensees. Luxfood Productions in contrast, was so much determined by its group's strategy that it got into conflict with the complementary organic quality aspects and was not able to develop an active marketing strategy for its organic product range. The findings thus support the propositions one and two of Clarke and Roome (1999 and in this volume) who claim that learning will most probably only take place when a strategy for learning exists and only regarding the issues that the strategy focuses on.

On a more general level, this case clearly shows that sustainable product solutions, like organic products, are still desperately in need of organisational actors that are both willing and able to develop markets (Dyllick et al., 1999). Partners along the product chain therefore should be chosen not only according to their ability to handle a certain product but also according to their possible interest in doing so. The power exerted on Luxfood Productions by Coop and Bio Suisse could not make Luxfood more open towards the organic products but rather made them bend as little as possible because the organisation had no strategic interest in the introduction of organic products. Thus we first conclude from this case that the exertion of power is not useful to make an unwilling partner be open for new experiences, let alone learn. We secondly conclude that although a power position is helpful to establish certain standards for sustainable products, there is more than one position on what a sustainable product can be. In a learning partnership the collaborating organisations should be open to mutual definitions, because the market development can also be inhibited by too rigid standards.

ACKNOWLEDGMENT

The author would like to thank the Swiss National Science Foundation that funded the research project under grant no. 5001-44663.

REFERENCES

Argyris, C. and D. Schön (1978) *Organizational learning. A theory of action perspective.* Reading, MA: Addison-Wesley.

Argyris, C. and D. Schön (1996) *Organizational learning II. Theory, Method and Practice.* Reading, MA: Addison-Wesley.

Arnold, S., C. Tanner and S. Wölfing Kast (1999) *Die Wirkung ausgewählter Kontextbedingungen auf das ökologisch nachhaltige Einkaufsverhalten: Resultate einer Tagebuchstudie.* [The influences of chosen context factors on ecological purchase behaviour: Results of a diary study.] Research Report from the Department of Psychology no. 1999-4. Bern: University of Bern.

Balzers-Graf, F.M. (1991) Steigbild und Kupferchloridkristallisation – Spiegel der Vitalaktivität von Lebensmitteln, in *Lebensmittelqualität – ganzheitliche Methoden und Konzepte* [Food quality – integral methods and concepts.] A. Meier-Ploeger and H. Vogtmann eds. Karlsruhe: Verlag C.F. Müller.

Bio Suisse Der Schweizer Biomarkt 1997 - 1999. [The Swiss Market for organic products 1997 - 1999.] Material delivered at the media conference; 2000 March 28; Bern.

Bio Suisse Hier essen Sie biologisch. [These places serve organic.], http://www.bio-suisse.ch/html/g_konsumenten_2.html retrieved on 2000 August 15.

Braune, P. and J. Franke (1992) Akzeptanz des Kantinenessens: ein innerbetriebliches Marketingproblem. [The acceptance of staff canteen food: an internal marketing problem.] *Zeitschrift für Organisation* (zfo), 1: 42-46.

Clarke, S. and N. Roome (1999) Sustainable Business: Learning-Action Networks as Organizational Assets. *Business Strategy and the Environment* 296-310.

Dyllick, T, F. Belz and U. Schneidewind (1999) Ökologie und Wettbewerbsfähigkeit. [Ecology and Competitive Advantage.] Hanser.

FAO (2000) Food safety and quality as affected by organic farming. Report of the 22nd FAO regional conference for Europe. July 24-28; Porto.

Fiol, C. and M. Lyles (1985) Organizational learning. *Organization Science* 5: 403-420.

Hartman, C., P. Hofman and E. Stafford (1999) Partnerships: A Path to Sustainability. *Business Strategy and the Environment* 255-266.

Hedberg, B. (1981) How organizations learn and un-learn, in *Handbook of Organizational Design 1.* B. Nystrom and W. Starbuck eds. New York.

Heller, T. (1998) *Bridging Compliance and Main Business Innovation: Managing Product and Process Development for a Sustainable Development.* Paper presented at the 7th GIN Conference; November 15-18; Rome.

Jungbluth, N. (1999). *Umweltfolgen des Nahrungsmittelkonsums: Beurteilung von Produktmerkmalen auf Grundlage einer modularen Ökobilanz.* [Ecological effects of food consumption: Evaluation of product characteristics based on a modular ecobalance method.] doctoral thesis, Zürich: Swiss Federal Institute of Technology (ETHZ).

Maier, S. (2001) *Strategische Herausforderungen bei der Einführung von Bioprodukten in der Schweizer Gastronomiebranche.* [Strategic challenges in the introduction of organic products to the Swiss catering industry.] unpublished doctoral thesis, St. Gallen: University of St. Gallen.

Maier, S., M. Finger and U. Haldimann (1999) *Sustainable development in the Swiss Food Sector: Restrictions and Strategies for Co-operations.* Paper presented at the Academy of Management Conference; August 9; Chicago.

Maier, S., M. Finger and U. Haldimann (1998) *Organisationale und interorganisationale Lernprozesse in Richtung Nachhaltigkeit im Bedürfnisfeld Ernährung. Forschungsdesign und erste Ergebnisse.* [Organisational and inter-organisational learning processes towards

sustainability. Research design and first results.] Idheap Working Paper No. 15. Chavannes-près-Renens: Idheap.
Matiaske, B. (1992) Europäisierung der Ernährungsstile: Genuss und/oder Gesundheit? in *Taste the difference. Europäische Ess- & Trinkkulturen im Wandel. Trends und Konsequenzen für Handel, Lebensmittelindustrie und Gastronomie.* Gottlieb Duttweiler ed.,Institut für wirtschaftliche und soziale Studien [European food and beverage culture under transformation. Trends and consequences for the retail trade, the food processing, and the catering industry.] Rüschlikon/Zürich: GDI Schrift Nr. 82.
Miles, M. and E. Weitzman (1994) Choosing computer programs for qualitative data analysis, in *Qualitative data analysis: an expanded sourcebook* M. Miles and M. Huberman eds. Thousand Oaks, CA: Sage Publications.
Minsch, J. and M. Mogalle (2000) *Wege zur nachhaltigen Ernährung – Schlussbericht des Integrierten Projektes Gesellschaft I.* [Paths to a more sustainable nutrition. Final report of the Integrated Project Society I: Sustainable Switzerland in the international context. Development of visions, strategies, and instruments for the need-field of nutrition]; IP – Discussion paper No. 5. St. Gallen: University of St. Gallen, IWÖ-HSG.
Mintzberg, H. and J.A. Waters (1985) Of Strategies, Deliberate and Emergent. *Strategic Management Journal* 6:257-272.
Nonaka, I. and H. Takeuchi (1995) *The knowledge-creating company.* Oxford.
Probst, G. and B. Büchel (1994) *Organisationales Lernen – Wettbewerbsvorteil der Zukunft.* [Organisational learning – Future competitive advantage.] Wiesbaden: Gabler.
Schneidewind, U. (1998) *Die Unternehmung als strukturpolitischer Akteur.* [The company as a structural political actor.] Marburg: Metropolis.
Tanner, C., S. Wölfing Kast, S. Arnold and K. Sätteli (1998) *Internale und externale Restriktionen und Ressourcen des ökologisch nachhaltigen Lebensmitteleinkaufs.* [Internal and external restrictions and resources of ecological food purchase.] Research Report from the Department of Psychology no. 1998-2. Bern: University of Bern.
UNCED (1993) *Agenda 21: the United Nations programme of action from Rio.* New York, NY: United Nations.
Villiger, A., R. Wüstenhagen and A. Meyer (2000) *Jenseits der Öko-Nische.* Basel/Boston/Berlin: Birkhäuser.
Welford, R. (1995) *Environmental strategy and sustainable development. The corporate challenge for the 21st century.* London/New York: Routledge.
Wüstenhagen, R. (1997) *Ökologie im Catering-Markt Schweiz: Branchenanalyse und Beurteilung ökologischer Wettbewerbsstrategien in der Gemeinschaftsgastronomie.* [Ecology in the Swiss catering market: strategic analysis of the catering industry and evaluation of ecological competitive strategies.] Discussion paper No. 45. St. Gallen: University of St. Gallen, IWÖ-HSG.
Yin, R.K. (1994) *Case study research: design and method.* Thousand Oaks, CA: Sage

Chapter 8

ENVIRONMENTAL ADAPTATION IN SUPPLIER-CUSTOMER RELATIONSHIPS

LOUISE CANNING

STUART HANMER-LLOYD

1. INTRODUCTION

This chapter focuses on partnerships between supplier and customer organisations which are used to reduce the environmental impact of the supply chain activities in which the companies are involved. A company might be able to pursue strategies for sustainable development without the contribution of or reliance on outside parties. However, companies do not operate in isolation of each other and some initiatives pursued by businesses affect customer or supplier organisations and will require their contribution. We therefore explore the processes followed by companies to introduce environmental adaptations where the support of exchange partners is required. By examining this process, the events and activities which result in environmental adaptation can be understood, as well as the factors which affect the progression of the process.

We present a brief review of literature relating to environmental supply chain co-operation before going on to summarise the IMP interaction framework (Håkansson, 1982) and to consider inter-firm adaptation. A case study research strategy is used where the relationship between two firms represents the case (or context) in which the environmental adaptation processes (the phenomena) are studied. We describe four case study relationships (two in the United Kingdom and two in Germany), featuring

T.J.N.M. de Bruijn and A. Tukker (eds.), Partnership and Leadership, 135–156.

five environmental adaptation processes. Data collection consisted of twenty-four in-depth personal interviews with managers from the supplier and customer companies who were involved in the processes to bring about the nominated environmental changes. Interviews covered the nature of the relationship with the specific supplier or customer company, company environmental policies and the managers' experience of and contribution to selected environmental adaptation processes.

We describe each relationship and type of environmental adaptation before going on to:

- analyse key features of the process of environmental adaptation;
- identify factors that affect the progression of the process;
- propose guidelines for managing the environmental adaptation process.

2. THEORETICAL FRAMEWORK

The literature reviewed here briefly considers reported forms of environmental supply chain co-operation before going on to present a summary of the IMP interaction framework and to discuss inter-firm adaptation. The IMP interaction framework is used to inform the description of supplier-customer relationships, thus providing the context for the investigation of environmental adaptation.

2.1 Environmental supply chain co-operation

Efforts to minimise environmental impacts shift beyond a company's own corporate boundaries as environmental performance is viewed as the responsibility of an entire supply chain rather than being the result of the behaviour of isolated actors (Shrivastava, 1995). A company might seek to reduce the impact of a product system either by analysing and subsequently trying to alter the life cycle of a product, or by concentrating its efforts on the implementation of a return loop system for products that it manufactures. In both cases, it is clear that the involvement of other companies in the supply chain is necessary. A number of authors have presented a variety of forms which environmental co-operation between firms might take, including:

- the exchange of material or product information (Berkhout, 1996; Cramer and Schot, 1993);
- the amendment to operational processes or products (Florida, 1996; Walton et al., 1998);
- the facilitation of product return systems (Meffert and Kirchgeorg, 1996).

However, in reporting such environmental co-operation little account is given of the way in which the adaptations are brought about, or the degree of ease with which they are introduced into existing relationships (Hass and Groenewegen, 1996). We therefore make use of literature associated with supplier-customer relationships and inter-firm co-operation in order to gain insight into firm behaviour which might be associated with environmental adaptation. Our interest is not to develop understanding of technical solutions to environmental problems, rather it is to shed light on the effective management of environmental supply chain co-operation.

2.2 IMP Interaction Framework[1]

The IMP interaction framework (Håkansson, 1982) provides researchers with a descriptive model which can be used to understand inter-firm behaviour in organisational markets. It is based on the premise that companies frequently enter into long term relationships in business markets where both supplier and customer organisations are active participants in the exchange processes. The framework (see Figure 1) consists of four elements which can be used to place environmental adaptations in the context of existing relationships. The four elements which comprise the framework are briefly presented in the following sections.

The Interaction Process
The interaction process reflects the form and types of exchanges that take place between companies as well as the development of these over time. A relationship can be presented in terms of the nature and role of a product/service supplied, the importance of either party to the other as well as the structure and nature of contact patterns between firms.

Interacting Parties
Interaction between companies is inevitably affected by the characteristics of the firms and by the individuals involved in a relationship. It is therefore appropriate to initially consider the nature of each company's activities and strategies, as these can lead to an understanding of the contribution of an exchange partner to a relationship and the expectations placed on that partner. Several individuals from both supplier and customer organisations are likely to have direct involvement in a relationship. Understanding of a relationship from the manager's perspective enables the researcher to take

[1]Work carried out by a group of European researchers in the early 1980s established the interactive nature of relationships in business markets and resulted in the development of a descriptive framework which can be used to explore industrial marketing and purchasing (IMP) relationships.

account of the inevitable variations in role, behaviour and perceptions of those that deal with partner organisations.

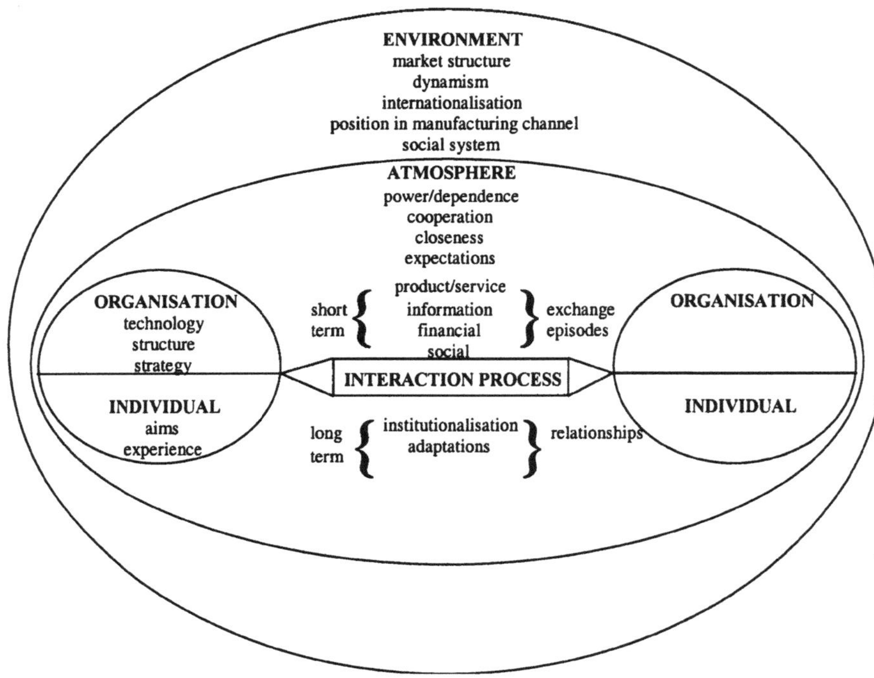

Figure 1. IMP Interaction Framework (Håkansson, 1982)

Relationship Atmosphere

Whilst a relationship might be described in terms of its 'structural' features (the interaction process and parties) the way in which companies or individuals deal with each other is determined by the 'emotional' atmosphere of the relationship. The intentions, negotiations and eventual agreements which are experienced and enacted by managers reflect the state of power/dependence, co-operation/conflict and social distance/closeness of a relationship.

Interaction Environment

It may not be possible to make use of the environment in which an inter-firm relationship exists in order to describe it. However an awareness of factors such as market structure, market dynamics and the position of interacting parties in the manufacturing channel can be helpful in understanding the actions of firms in relation to each other.

Although the IMP interaction framework provides a useful tool for the description of interactive supplier-customer relationships, it can be difficult to use given the range of variables that it includes. In order to deal with this and bearing in mind that the focus of our investigation centres on the process of environmental adaptation, our description of the selected relationships concentrates on those elements likely to be associated with adaptation behaviour and are shown in Figure 2.

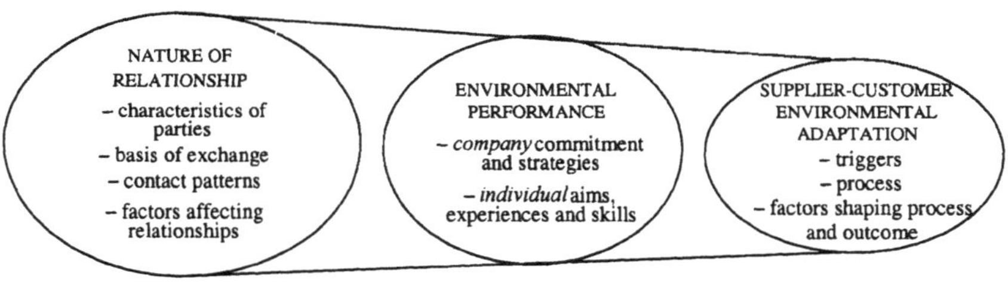

Figure 2. Inter-Firm relationship and adaptation behaviour

2.3 Adaptations in supplier-customer relationships

Adaptations can be attributed to two main causes, namely the desire by companies to improve the exchange relationship with a partner organisation and the need to respond to market dynamics. Supplier-customer relationships are characterised by firms that have varying resources and capabilities as well as efficiency and profitability criteria (Hallen et al., 1993). Environmental adaptations will occur as companies in an exchange relationship try to better align their activities in order to satisfy the environmental performance criteria of the respective parties and ensure the efficient use of resources (Florida, 1996). Adaptations will also occur in business relationships as the exchange parties respond to the changing business conditions to which they are exposed (Hallen et al., 1991). For example, the introduction (or anticipation) of environmental legislation continues to change the behaviour of individual firms and also leads to adaptations in supplier-customer relationships.

Where either company in a relationship pursues adaptations in the elements exchanged or the operational processes or technologies used by the other party, then the involvement of that other party in the environmental adaptation process is necessary. This means that before the eventual change can be made, some degree of co-ordinated action, or indeed co-operation has to occur between companies in order to agree the nature of the necessary change and the contribution of both parties to facilitating the adaptation. Such co-ordination or co-operation would be expected to include "a repetitive series of negotiation, commitment and execution stages" (Ring and Van de Ven, 1994). Ring and Van de Ven's interest lies in the development of new relationships. However, the notion that this developmental process is made up of a series of repeated episodes which involve negotiation, agreement and action is central to the process of environmental adaptation in supplier-customer relationships.

3. THE CASE STUDIES

The findings presented here describe each supplier-customer relationship and the environmental adaptations which featured in these relationships.

3.1 Case study number one

This case study involves a supplier of telecommunications equipment and its distributor in the United Kingdom. The environmental adaptation which features in the relationship comprises a product return and remarketing system for used telecommunications equipment. Both parties recognised the need to develop such a scheme in response to legislative and market pressures. Whilst a trial scheme was successfully completed, at the time of the study, strategic issues associated with the handling of complex equipment, each party's involvement in the used equipment market and the identification of markets for second-hand equipment had yet to be addressed.

The Relationship

The relationship between the telecommunications equipment manufacturer and distributor in the United Kingdom is important for both companies. The distributor is one of the manufacturer's largest customers world-wide and acts as the company's sole distributor in the United Kingdom. The distributor does not represent any other manufacturers of similar equipment. The relationship between the supplier and distributor is very close and fairly well integrated as indicated by the following:

- Sales and marketing activity to end customers is split between the manufacturer's sales team and the distributor's product and channel marketing team.
- The companies operate a joint business development group.

- The distributor is closely involved in supplier new product development activity.
- Job exchange schemes exist between the two organisations.
- Interpersonal contact is frequent at a variety of levels.
- Growing competition as a result of market deregulation has affected both organisations.

Much of the daily contact between the two companies involves the manufacturer's account team and the distributors supply and product marketing managers. The focus of these commercial discussions includes the consideration of new sales opportunities or competitive threats, the review of supply requirements, price adjustments and the provision of promotional and training activities. The relationship can be described as close, although some tension does occasionally exist associated with reluctance to share cost and pricing information.

Company Environmental Policies
Both companies are actively engaged in trying to reduce the environmental impact of their products and processes and have won commendations for their efforts. In both organisations, expectations of employee contribution to environmental initiatives is highest amongst those involved in product development and operational activities. Managers responsible for the administration of the distribution relationship have had little direct involvement in their respective company's environmental initiatives.

The Adaptation
The environmental adaptation which both companies sought to introduce, was a product return and resale system for used telecommunications equipment. This meant that the primary responsibility for facilitating the introduction of such a system lay with the relationship management teams in both companies.

3.2 Case study number two

This case study involves a supplier of cables and a telecommunications operator in the United Kingdom. The environmental adaptations which featured in the relationship comprise packaging redesign as well as return systems. Whilst the customer driven initiative now operates successfully (in spite of initial difficulties), the supplier's effort did not succeed.

The Relationship

The relationship between the cable supplier and the telecommunications service provider is important for both companies. The cable manufacturer is the single source of supply used by the telecommunications company to connect its exchange network. The telecommunications company accounts for 30% of the cable supplier's turnover in the United Kingdom. The relationship between the two companies is quite close, as witnessed by:

- The willingness of the telecommunications company to use a single sourcing strategy for the supply of a product which is key to its service operations.
- The introduction of a cross-functional customer account team (including development, production, technical and sales personnel) by the supplier.
- Frequent interpersonal contact.

Much of the interpersonal contact between the two companies is between the manufacturer's account manager and the customer's supply manager. The meetings are used to discuss the commercial administration of ongoing contracts. Efficiency improvements expected of the telecommunications company means that ways in which the cable company might reduce the costs of supply to its customer represent a significant part of the regular meetings between the companies.

Company Environmental Policies

The hazardous nature of the cable manufacturer's operations means that its manufacturing processes are closely scrutinised by external bodies. Efforts to improve the company's environmental performance have centred on manufacturing activities, for which the company has gained recognition. Commercial managers in the customer account team have little exposure or expected contribution to the company's environmental performance. The telecommunications company actively seeks to reduce the environmental impact of its own operations as well as the supply chain of which it is a part. Supply managers are expected to consider a supplier's environmental policy as well as the environmental impact of items purchased.

The Adaptation

Both companies have sought to introduce packaging return systems. Whilst the batten return system initiated by the supplier failed, the drum return system initiated by the supply manager was successfully introduced. The companies are also currently investigating packaging design improvements.

3.3 Case study number three

This case study involves a plastics supplier and an IT company in Germany. The environmental adaptation which was initiated by the IT company comprises the development of a return and recycling system for used plastic retrieved from their company's products. The IT company has now become a supplier of used material feedstock to the plastics company.

The Relationship

The relationship between the German plastics manufacturer and the German IT company is not important for either organisation in terms of its financial value or the function of the plastic, whose primary purpose is to contain and protect the electronic systems designed by the IT company. The relationship is considered by both parties to be co-operative and is based on the manufacturer's expertise in engineering plastics which is used to guide the IT company as well as subcontractors in the supply chain. The degree of interdependence exhibited in cases one and two does not exist due to:

- Low volume and financial value of exchanges.
- Sporadic interpersonal contact.
- Injection moulding of plastics by a third party prior to final product assembly by the IT company.

Much of the contact between the two companies involves the supplier's technical sales manager and the IT company's designers and supply centre managers. The exchanges centre around new product designs and material requirements which are carried out by telephone or written communication. The subcontracting of non-core activities by the IT company means that as well as dealing with this organisation, the plastics manufacturer manages relationships with the subcontractors in order to secure business with the IT company.

Company Environmental Policies

The plastics manufacturer places a high level of importance on environmental health and safety considerations both in terms of its internal processes as well as products supplied to customers. The company has a plastics recycling unit which is designed to support the sale of products to customers and sales managers are required to offer a range of plastics to customers, including those produced from used material. The IT company considers itself to be a leader in its industry in terms of its efforts to minimise the impact of its activities on the environment. The company's performance has been commended and the environmental design of its products has enabled it to compete successfully against major IT companies.

The company's designers are expected to agree material specifications with suppliers which conform to the company's internal environmental product design guidelines. Supply managers are responsible for the commercial negotiation of such agreements.

The Adaptation
Unlike most of its competitors, the IT company has established its own recycling centre in order to manage the reconditioning of used equipment and the reclamation of materials from obsolete items. Efforts to eliminate the disposal of plastics via landfill resulted in the investigation of material recycling with the plastics manufacturer.

3.4 Case study number four

This case study involves a supplier of label technology and the IT company from case number three. The adaptation also involves some limited contribution from the plastics supplier that featured in case number three. The adaptation which was triggered by the IT company, comprises the development of labelling materials which are compatible with plastics, thus making their removal before the recycling of used plastics unnecessary.

The Relationship
The relationship between the American owned identification systems company and the German IT company is not significant in terms of the product function or financial value of the exchanges between the two organisations. The identification systems company sells base materials to third parties who then convert these into labels for use by the IT company. The function of the product and its indirect supply via third parties means that a close relationship between the supplier and the IT company does not exist. Designers within the IT company make intermittent use of the guidance available from the supplier in the design and selection of label systems. Such exchanges between the supplier's sales managers and the IT company's designers represents the main contact between the two organisations.

Company Environmental Policies
The identification systems company has operated a formal environmental policy since the 1970s and has won international recognition for its efforts to reduce the environmental impact of its operations and products. The primary responsibility of sales managers is to relate customer demands for less environmentally damaging products to technical and product development

departments. The IT company's environmental policies are described in case number three.

The Adaptation

The fact that the IT company is directly involved in the recycling and reclamation of used equipment exposed the company to difficulties (and therefore increased costs) associated with the removal of labels during the preparation of plastics for recycling. The desire to reduce such costs has resulted in the development of a label which is compatible with plastics and does not need to be removed before plastics are recycled.

4. ANALYSIS OF THE ENVIRONMENTAL ADAPTATION PROCESS

Having provided an overview of the four case study relationships and the adaptations that feature in them, we can now turn our attention to the processes which lead to the selected environmental adaptations. We consider, in particular:

- the factors which stimulate the pursuit of environmental adaptations;
- the nature of the adaptation processes;
- the factors which facilitate and hinder the progression and successful conclusion of environmental co-operation.

4.1 Factors motivating the pursuit of adaptations

The adaptations introduced into the four relationships resulted from efforts by companies to respond to elements of uncertainty and to achieve cost improvements either within one party's internal activities or as part of the relationship with the exchange partner. The drivers associated with uncertainty and cost reduction are now discussed.

Uncertainty

Uncertainty resulting from environmental legislation and competitive market dynamics represents a key contributor to efforts by companies to seek to introduce adaptations into relationships with selected supplier and customer companies. Relationships with other organisations present companies with a framework in which they can seek to respond to the opportunities or problems resulting from such uncertainty (Hallen et al., 1987). In the same way that environmental legislation has represented a primary stimulant for the development of environmental adaptation within companies, the introduction or anticipation of legislation aimed at the electronics industry led the equipment supplier and distributor in case number one to try and deal with the resulting uncertainty via their relationship with each other.

Whilst legislation might trigger the need to develop appropriate strategies amongst managers with distinct environmental responsibilities, for those managers involved in the commercial administration of exchange relationships, uncertainty resulting from competitive market dynamics represented a stronger motivation for seeking to pursue changes with a partner organisation. In case number one, the growth in the availability of used equipment was perceived as a potential threat to new equipment sales by the 'commercial' managers. Despite the different factors motivating the commercial and environmental managers to pursue adaptations in this relationship, a common objective did exist, namely to control the flow of used telecommunications equipment and to harness the value attached to it.

Cost

Findings from the case studies support the view that adaptations are also stimulated by the desire of parties to achieve cost reductions (Makower, 1993). In cases three and four efforts by the IT company to reduce its own internal waste/recycling costs resulted in the pursuit of plastics/labels initiatives with its suppliers. In case number two, efforts by the cables supplier and telecommunications operator to reduce the cost of packaging and disposal led to the investigation of return systems and the redesign of packaging. Whilst the IT company's interest was directly associated with the organisation's environmental product strategy, the packaging projects pursued in case number two were primarily attributable to general efforts to improve cost performance. Nevertheless, similarities in materials management and environmental objectives of less resource use and waste (Steger, 1993) means that these might equally be considered as environmental adaptations. The focus on costs in interactions between managers responsible for relationship administration means that this area of combined environmental and cost improvements is likely to appeal to and perhaps be the responsibility of 'boundary personnel' such as supply and customer account managers.

4.2 The process of adaptation

All of the adaptations described by managers in the case study research feature processes in which the contribution of both parties is necessary in order that progress towards the eventual adaptation be made. The findings therefore confirm the importance of inter-firm co-operation in achieving environmental improvements (Aulinger, 1996; Cramer and Schot, 1993; Green et al., 1996; Steger, 1996). This co-operation consists of a co-ordinated series of actions involving negotiation (or information exchange), commitment to and the execution of tasks.

Negotiation, Commitment and Execution

Information exchange has previously been used to indicate environmental supply chain co-operation, although it appears to be restricted to impersonal channels (Cramer and Schot, 1993; Steger, 1996). The research findings establish the importance of interpersonal communication between company representatives in facilitating the progression towards an eventual adaptation output. This form of information exchange can encourage participation in and acceptance of the eventual output of adaptation processes. Interpersonal communication which occurred in the selected adaptations served the following purposes:

- *the familiarisation of managers;*

  Adaptations which either drew upon the environmental strategies of companies or exchange relationships between firms required the involvement of managers who were unfamiliar with either technical aspects of the adaptation or the relationship with the partner organisation. The initial communication episodes therefore serve to introduce managers to these issues and the relationship.

- *shared understanding of the nature of the problem;*

  Irrespective of whether companies had direct experience of dealing with the particular issue for which a solution is being sought, a common awareness of and agreement on the nature of the problem being faced acted as the basis for the development of an eventual adaptation output.

- *interpretation and agreement of issues to be addressed;*

  Agreement on the nature of the problem as perceived by managers from each company enabled the parties to establish areas that required consideration from company and functional perspectives.

- *presentation of expertise/resources to which either party has access.*

  Whilst managers might have agreed on the nature of the problem and issues to be addressed, the ability to do this necessitated access to specific expertise or resources. Communication episodes between companies allowed parties to present the expertise which had been developed and which could be used to either guide or contribute to the adaptation process.

Such communication activities are important where the problem which stimulates the adaptation process covers issues which previously might not have been part of the relationship between two companies. As a result, managers normally responsible for the administration of the relationship might be required to broaden the focus of their exchange activities with the partner organisation or managers not normally involved with a supplier or customer company might be expected to participate in exchanges with that organisation.

The research findings suggest that communication episodes are clearly used to establish the basis for progression of the adaptation process and to agree on the activities to be carried out by each party. Extensive bargaining, where one party might seek to coerce the other to assume responsibility for certain tasks, did not appear to dominate the information exchange episodes. Where the nature of the problem indicated clear areas of responsibility and each party in the process had a particular expertise which could be used to facilitate the progression of the process, then little negotiation regarding the allocation of tasks and therefore costs occurred.

Whilst companies might enter into formal contractual arrangements with each other, commitments can equally be made via informal agreements between exchange partners. This was the situation in this investigation of environmental supply chain co-operation. Agreements reached on the allocation of tasks during the adaptation process and the commitment to pursue selected activities which resulted in the investment of resources by companies, were not underpinned by formal contracts between the supplier and customer companies.

Although adaptation processes might involve a series of negotiation, commitment and execution stages where companies interact with each other, the actual execution of tasks to which company representatives commit themselves then necessitates co-operative processes within an organisation to enable the execution of those previously agreed actions. For example, in case number two, the supply manager sought to ensure that operations managers made packaging available for return to the cable supplier. In case number three, the materials recycling manager employed by the plastics manufacturer sought the approval of the customer account manager before pursuing plastics recycling with the IT company. In case number four, the supplier's product environmental affairs and customer account managers presented a business case in support of plastics compatible labels before the development of the new product could be pursued.

This means that those involved in the adaptation process with a partner organisation and who commit to executing certain tasks need to possess resources which enable them to carry out the actions to which they have agreed, which in turn allows the adaptation process to proceed.

Form of the Adaptation Process

The risk or uncertainty associated with the adaptation being sought inevitably affects both the number of repeated negotiation, commitment and execution phases, as well as the range of people involved in those processes. For example, in case number two, the packaging return scheme initiated by the telecommunications company represented a system which was already familiar to the manufacturer and the customer's supply manager. The initial introduction of this logistics system therefore required limited exchange episodes between the companies and only involved the manufacturer's account manager and the telecommunication company's supply manager

In case number one, the equipment supplier and the distributor had offered product return and resale to customers on previous occasions. However, the combined significance of the need to deal with impending legislation and the degree of uncertainty surrounding the new scheme and eventual resale of used equipment necessitated a more cautious and extended series of exchanges between the companies for the new return scheme. Whilst environmental and recycling managers from both companies were initially involved in discussions, the fact that the return and resale system involved primarily logistical issues meant that commercial managers represented the key actors in this adaptation process.

Changes to materials, products and packaging specifications which are central to the activities of either or both partner organisations represent a high degree of uncertainty, requiring a considered and extended series of stages/milestones before adaptations are eventually introduced. This involves more intense interaction between the companies and investment in trials, prototype tests and investigations by the organisations. For example, in case number two, efforts to change a key packaging design required the development and testing of prototypes by the cable manufacturer before submission to the telecommunications company for its own internal testing and approval. Although the design and testing of the packaging in each company was carried out by technical managers, the process was co-ordinated by the cable manufacturer's account manager and the customer's supply manager.

In cases three and four, the supply and recycling of used plastic and the development of plastic compatible labels necessitated the discussion of specifications, material requirements, the development and supply of prototypes as well as trials to enable the adaptation processes to proceed. The technical emphasis of such adaptations inevitably meant that those managers responsible for product design/specification, performance and recycling assumed primary responsibility for contributing to the adaptation processes with partner organisations. Relationship managers were however either involved in meetings or informed of progress as the adaptation processes proceeded and assumed primary responsibility once the emphasis shifted to commercial issues.

4.3 Factors affecting progression of the process

The adaptations featured in the four case studies involve a series of repeated inter-firm exchanges and of tasks executed by individual managers which enabled the processes to proceed to their eventual conclusion. Co-ordinating activities between companies requires the contribution of managers from both supplier and customer organisations. The successful conclusion of the adaptation processes and the nature of the adaptation outputs was affected by the behaviour and direct experiences of those managers involved in exchanges with supplier or customer organisations.

Manager Behaviour

The research shows that the behaviour of managers involved in the adaptation is influenced by company strategies as well as the personal aims of individual managers themselves. For example, in case number three both the plastics manufacturer and the IT company had clear strategies which supported the marketing of new and used or recycled material and equipment. The availability of used or recycled items was not considered to be a threat to new sales, and sales managers were expected to make both available to customers. In case number one however, the telecommunications equipment supplier and the distributor had yet to develop corporate policies which supported the resale of used equipment in the United Kingdom. This absence meant that sales and marketing managers in both companies had little incentive to actively support efforts to develop product return and resale schemes for used equipment.

During the adaptation processes, manager behaviour was also determined by the degree of interest an individual had in seeking to facilitate the process. Where a change being pursued was closely linked to the expectations placed on a manager and to that person's normal work activities, then that manager had a higher degree of interest in and contribution to (or influence over) the progression of the process. For example, in cases three and four, those managers with environmental or recycling responsibilities were key contributors to the process. The managers were involved in inter-firm exchanges and in executing agreed commitments (such as co-ordinating the activities within their own companies). In case study number two, the primary responsibility of the telecommunications company's supply manager was to ensure continuity of supply at minimum cost. This motivated the manager to initiate a packaging return scheme and to seek to resolve difficulties with the initial operation of the scheme. Such behaviour is contrasted with the efforts by the cable manufacturer to introduce a return system for packaging used to supply other products. The ability of the supplier to operate the scheme required the support of the manager responsible for handling the supplies of the particular product associated with this packaging. In this instance it can be concluded that the benefits of such a scheme were less apparent to the supply managers as their facilitation in operating the return scheme did not occur.

Manager Experience

In addition to the adaptation process being affected by the behaviour of managers, its progression is also determined by the experiences of those managers involved. Such experience contributes to:

- the level of satisfaction that managers attach to the outcome of exchange episodes with partner organisations;
- the perceived trustworthiness of exchange partners;
- the perceived commitment of supplier or customer companies to the introduction of the adaptations.

Exchange Outcomes

Where exchanges between companies involve some negotiation of how to pursue a specific adaptation or the conditions for its eventual operation, management satisfaction is associated with another party's efforts to shape either the process or outcome. For example, in case number one, dissatisfaction emerged amongst the equipment supplier's managers as a result of the distributor's efforts to determine markets for the sale of used equipment. Whilst this does not halt the progression of the scheme, it will influence its design and also affect the supplier's relationships with distributors in other markets.

Trustworthiness

Managers judge the trustworthiness of a supplier or customer company based on their credibility (Mayer et al., 1995) and reliability (Doney and Cannon, 1997; Ganesan, 1994). Where companies seek to resolve a new problem, or managers not normally involved in the exchange relationship become involved in the environmental adaptation process, then an assessment of the credibility and expertise of the supplier or customer company is made during the initial inter-firm exchanges. For example, in cases three and four, managers newly introduced to the respective exchange relationships formed a judgement of the expertise and resources available to the partner organisation before opting to support the environmental adaptation process.

In addition to perceived trustworthiness being associated with the capabilities of a company, managers also make judgements based on the extent to which a partner's word can be relied upon. In this research, reliability was associated by managers with the timeliness of the execution of actions to which companies had committed during inter-firm exchanges. For example, in case number three the plastics manufacturer perceived a high level of consistency between the agreement to and execution of actions by the IT company. This acted as an indicator of the IT company's interest in plastics recycling and facilitated the progression towards an eventual recycling system.

Commitment

Commitment is typically associated with inter-firm relationships, yet it might equally be applied to adaptations also. Judging the degree of effort which a partner expends to facilitate the adaptation process is significant, as this perceived commitment in turn shapes the effort (or commitment) of the other company (Anderson and Weitz, 1992). In this research, the timeliness in executing agreed actions appeared to act as an indicator of a partner's commitment to a specific adaptation process and eventual output. In case number three, the timeliness of the execution of agreements by the IT company acted as an indicator to the plastics manufacturer of the customer company's commitment to plastics recycling. By way of contrast, in case number one the delays in the supply of pricing information led the distributor to question the commitment of the telecommunications equipment supplier to the development of a product return and resale system.

In addition to the timely execution of tasks, continuity in the structure of contact patterns in inter-firm exchanges throughout the lifetime of projects was noted by suppliers in cases three and four as facilitating the adaptation process. As well as ensuring that managers had direct access to relevant people, which enabled the speedy resolution of issues and eliminated the distortion of information, such continuity of company representatives involved in inter-firm exchanges was perceived by the supplier's managers to signal the IT company's commitment to the adaptations being sought. Such a perception is contrasted with case number one where the distributor's managers concluded a lack of commitment on the part of the equipment supplier due to the absence of the supplier's recycling managers beyond the initial discussion regarding a return and resale system for used telecommunications equipment.

5. CONCLUSIONS AND MANAGEMENT GUIDELINES

The environmental adaptation process involves both parties and the decision of either party to support the process is to some degree determined by an exchange partner's own interest in the eventual adaptation output as well as the perceived commitment, credibility and reliability of the other party. This perception is formed through involvement in the adaptation process and through experience of the exchange partner. From the analysis it can be seen that a number of factors can hinder or facilitate the progression of the adaptation process. These are presented in Table 1. The guidelines which follow suggest how firms might deal with these factors both within their own organisation and in dealing with their adaptation partners.

Table 1. Factors affecting progression of adaptation process

Barriers	**Facilitators**
– the absence of support from senior management limits the willingness of managers to assist the process of adaptation – unless there are clearly identified benefits resulting from the adaptations, manager effort to help the introduction of changes will be limited – managers are unlikely to support the introduction of adaptations which undermine their ability to meet their own performance targets – interest in self rather than mutual gain can reduce partner motivation to co-operate – companies are unlikely to encourage adaptations which are perceived to place a key element of their business activity at risk – where high risk adaptations are being pursued, an extended adaptation process is required – discontinuity of managers involved in information exchanges and delays in carrying out tasks can suggest a lack of interest in the adaptation and so demotivate an exchange partner – slow task execution can hamper partner actions and progress	– managers are more likely to support changes where there is consistency between their performance targets and their expected contribution to the adaptation process – support can be won where benefits resulting from changes can be clearly attributed to a manager's own performance, a company or the relationship between two organisations – information exchange helps to clarify interests, needs, sharing of tasks and conditions for progress – prior experience can reduce the extent of information exchange needed – previous problem-solving experience and success can enhance partner confidence in a company's capabilities – trials can reduce the perceived uncertainty associated with high risk adaptations – continuity in contact patterns facilitates information exchange, – the timeliness of task execution enables process to progress – resource investment by both parties fosters mutual process

Intracompany: Prominence of expected employee contribution to company environmental performance

Clear indications for employees are needed (especially those involved in the management of a supplier or customer relationship) regarding the relative importance of a company's environmental performance compared to other performance targets which employees are expected to help achieve. Where reduced environmental impact affects a company's business strategy then clear statements and proposals are needed detailing how the combined pursuit of environmental and business performance targets will be achieved. Such statements require the support of senior management and need to be reflected in the targets against which middle managers' performance is measured. Without the support of senior management or rewards for achieving environmental and other business objectives, the process of environmental adaptation will be hampered.

Inter-company: Information exchange and negotiation

The ability of companies to introduce a change depends on the willingness of exchange partners to facilitate the process and accept the adaptation being proposed. A company must therefore identify ways to ensure a partner's support. This requires that a company considers the interests of the exchange partner and that the company reflects on the extent to which it might be viewed as a credible, trustworthy and committed contributor to the adaptation process by its exchange partner. To do this a company must firstly recognise that managers involved in the adaptation process are inevitably influenced by their specific managerial responsibilities. A company should identify ways in which the eventual adaptation might be seen to support the activities of the representatives that they normally deal with in the supplier or customer company. Alternatively a company could target other managers whose responsibilities might result in the adaptation being supported.

Secondly, a company needs to persuade an exchange partner of its expertise and experience in dealing with the particular environmental adaptation which is at the centre of the co-operative process between the two companies. To do this managers involved in exchanges with a partner organisation must possess a range of expertise that reflects the adaptation, enabling the partner's queries to be clarified, thus reassuring the company representatives about the organisation's problem solving capabilities. Although complex adaptations will require the involvement of managers not normally associated with an exchange partner, the contribution of relationship managers should not be ignored, even in technical information exchange episodes. Supplier and customer facing managers will be more familiar with an exchange partner and can therefore ensure that contact with appropriate managers in the supplier or customer company is made.

Thirdly, a company should demonstrate its trustworthiness and commitment by ensuring the timely execution of tasks to which it has committed and by maintaining continuity in the contact patterns with an exchange partner during the adaptation process. Timeliness in the completion of tasks is particularly important where the partner company is dependent on these tasks being completed so it can perform activities to which it has agreed as part of the adaptation process.

The adaptations investigated in this study exemplify the increase in environmental co-operation predicted by Cramer and Schot (1993) and show that this can be more than impersonal information exchange. This research expands understanding of environmental co-operation (or adaptation) to consider the process by which this is achieved rather than simply noting the form that it takes. By studying the process, the activities and events associated with the introduction of environmental adaptations have been identified and factors which might affect the progression of the process have been established. The difficulties recounted in some of the environmental adaptation processes exemplify the suggestion by Hass and Groenewegen (1996) that environmental improvements can not necessarily be seamlessly introduced into business relationships. Whilst some of these difficulties might derive from technical problems, the progression of the process is shown at times to be conditioned by the relationship between the two companies as well as the behaviour and experience of those involved in the process. So trying to reduce the environmental impact of company activities not only requires the consideration of the optimum technical solution but also necessitates the effective management of the co-operative process between firms. As can be deduced from the guidelines, a company should not only consider what it can contribute to the adaptation process, but ways in which it can facilitate the involvement of the other organisation.

REFERENCES

Anderson, E. and B. Weitz (1992) The use of pledges to build and sustain commitment in distribution channels. *Journal of Marketing Research* 29, 1: 18-34.

Aulinger, A. (1996) *(Ko-)Operation Ökologie, Kooperation im Rahmen ölologischer Unternehmenspolitik* (Ecological co-operation: co-operation as part of environmental business policies). Marburg: Metropolis-Verlag.

Berkhout, F. (1996) Life-cycle assessment in large firms. *Businesss Strategy and the Environment* 5, 3:145-155.

Cramer, J. and J. Schot (1993) Environmental co-makership amongst firms as a cornerstone in the striving for sustainable development, in *Environmental Strategies for Industry:*

International Perspectives on Research Needs and Policy Implications, J. Schot, K. Fischer, eds. Washington DC: Island Press.

Doney, P.M. and J.P. Cannon (1997) An examination of the nature of trust in buyer-seller relationships. *Journal of Marketing* 61, 2: 35-51.

Florida, R. (1996) Lean and green: the move to environmentally conscious manufacturing. *California Management Review* 39, 1: 80-105.

Ganesan, S. (1994) Determinants of long-term orientation in buyer-seller relationships. *Journal of Marketing* 58, 2: 1-19.

Green, K., B. Morton and S. New (1996) Purchasing and environmental management: interaction, policies and opportunities. *Business Strategy and the Environment* 5, 3: 188-197.

Håkansson, H. (ed.) (1982) *International Marketing and Purchasing of Industrial Goods, An Interaction Approach*, New York: John Wiley and Sons.

Hallen, L., J. Johanson and N. Seyed-Mohamed (1987) Relationship strength and stability in international and domestic industrial marketing. *Industrial Marketing and Purchasing* 2, 3: 22-37.

Hallen, L., J. Johanson and N. Seyed-Mohamed (1991) Inter-firm adaptation in business relationships. *Journal of Marketing* 55, 2: 29-37.

Hallen, L., J. Johanson and N. Seyed-Mohamed (1993) Dyadic business relationships and customer technologies, *Journal of Business to Business Marketing*, 1, 4: 63-90.

Hass, J.L. and P. Groenewegen (1996) Strategic co-operation and life cycle analysis, in *The Greening of Industry Resource Guide and Bibliography* P. Groenewegen, K. Fischer, E. Jenkins, J. Schot eds. Washington DC: Island Press.

Makower, J. (1993) *The E-Factor*. New York: Tilden Press.

Mayer, R.C., J.H. Davis and F.D. Schoorman (1995) An integrative model of organisational trust. *Academy of Management Review* 20: 709-734.

Meffert, H. and M. Kirchgeorg (1996) *Strategies for the implementation of closed loop value chains - an empirical study of German manufacturers*. 5th Greening of Industry Network Conference; November 24-27; Heidelberg.

Ring, P.A. and A.H. Van de Ven (1994) Developmental processes of co-operative interorganisational relationships. *Academy of Management Review* 19, 1: 90-118.

Shrivastava, P. (1995) The role of corporations in achieving ecological sustainability. *Academy of Management Review* 20, 4: 936-960.

Steger, U. (1993) The greening of the boardroom: how companies are dealing with environmental issues, in *Environmental Strategies for Industry: International Perspectives on Research Needs and Policy Implications*, J. Schot, K. Fischer eds. Washington DC: Island Press.

Steger, U. (1996) Managerial issues in closing the loop. *Business Strategy and the Environment* 5, 4: 252-268.

Walton, S.V., R.B. Handfield and S.A. Melnyk (1998) The green supply chain: integrating suppliers into environmental management processes. *International Journal of Purchasing and Materials Management* 34, 2: 2-11.

Chapter 9

SUNNY DEVELOPMENTS

The Diffusion of Photovoltaic Technologies in The Netherlands

JOANNEKE KRUIJSEN

1. INTRODUCTION

Sustainable development is widely assumed to imply changes in our patterns of behaviour and consumption. An increase in the use of renewable energy would allow the world to make these changes. Part of the renewable energy supply mix consists of photovoltaic solar energy. The challenge of increasing the use of photovoltaic technologies is not a technological challenge one only: it is also a social challenge. The new technology needs to be accepted by its social environment before it can develop into a substantial part of the sustainable or renewable energy supply mix.

The process of technology diffusion in which a newly developed technology is accepted by its social environment formed the object of the author's study on photovoltaic (PV) technology diffusion (Kruijsen, 1999). Both the Dutch and the Californian PV sectors were investigated. Two focal points are important in the study: the role of diffusion networks and the role of product innovation. Most of the approaches for the implementation of new technologies into product innovations focus only on the level of an individual organisation. The study presented here, however, also includes the level of the inter-organisational network.

T.J.N.M. de Bruijn and A. Tukker (eds.), Partnership and Leadership, 157–175.

The collaboration within the Dutch diffusion network to advance PV diffusion is elaborated in this chapter[1]. The window frame or analytical framework through which we look into the Dutch situation is firstly described, followed by the observations as obtained while looking through that window. On the basis of these observations, several strategies for PV diffusion are given.

2. PARTNERSHIPS IN DIFFUSION NETWORKS

Analysing the collaboration and partnerships for PV diffusion requires a clear analysis framework. For this purpose, an analytical model was set up. The basic assumption for this model is that technology diffusion is defined as embedding a new technology into society by dissemination and application of this technology.

2.1 Dissemination

Dissemination into society refers to both the spread of information about the technology (the software) and the spread of the physical embodiment of the technology (the hardware). Dissemination takes place in diffusion networks: groups of interrelated social units, individuals or organisations, focused on technology diffusion. Therefore, the dissemination of technology can be analysed using a network approach. This network approach is characterised by four network functions and by two network interactions (also see the upper part of Figure 1).

The four network functions are: technology development, product development, use and stimulation (based on the work by Håkansson, 1985; Von Hippel, 1988 and Lundgren, 1995). Technology development is the function of those actors in the diffusion network who innovate new photovoltaic technologies, improve existing photovoltaic technologies, design, produce and/or test components for photovoltaic solar energy systems. Product development is the bridge between technology development and the use of the applied technologies. Product developers apply the photovoltaic technologies when they design, produce, and/or supply product systems with these technologies. Use is the function of those actors who use the product systems with the photovoltaic technologies. And finally, stimulation is the function of those actors who support and enable the functioning of the other three functions in the network. It is very well

[1]The Dutch situation is further described in a paper by Kruijsen, Bouwmans and Weijnen (1998) that was the basis for this chapter. The Californian situation is described in a paper by Kruijsen and Bouwmans (1998).

possible that an actor, an organisation in the diffusion network, can perform more than one function, e.g. a technology developer has a department for the production of solar cells and also has a department for the development of solar systems.

The two network strategies are: direct and indirect coupling. Direct coupling means direct interactions between two or more functions to exchange knowledge, experience, technology, money, etc. Indirect coupling means the interactions of one actor with two or more other actors and thus coupling these actors indirectly, e.g. when there is a lack of awareness of the possibility of coupling. Examples of indirect coupling are a consultant who brings a technology developer and a product developer together for interaction on the exchange of technological solutions, and a consumer organisation that provides a user with information on products from different product developers.

2.2 Application

Application takes place during the product development process. In this study we focus on a product as a physical result of a manufacturing process that performs a specific function by combining the need for that function with a technology to fulfil that need. A product can be one object or a configuration of objects: a product system. The application of a technology can be analysed using a product development approach. This product development approach is characterised by four process blocks and by two external sources for the product development process (also see the lower part of Figure 1).

The four process blocks are based on the 'basic cycle of design' (Roozenburg and Eekels, 1995): analysis, synthesis, simulation and evaluation. The basic cycle is an iterative process in which each block is used at least once when solving a design problem. The importance of the analysis block consists in collecting data to obtain insight into the problematic situation and to make a list of requirements for the new product. The essence of the synthesis block is the creation of a complete whole out of separate elements. The complete whole will be a provisional design of the new situation that might fulfil the future needs and wants. The essence of the simulation block is to assess which properties of performance the provisional design might have: is the provisional design capable of fulfilling future needs and wants? Finally, the importance of the evaluation block consists in identifying the difference between the desired properties and the expected properties and in deciding whether the differences are too large or acceptable. If the differences are too large, the development process has not ended yet: new blocks of synthesis, simulation or even analysis are needed.

The two external sources are user wants and technological options. In the study, we considered product development to be the process of combining the needs of the user with the technological options to supply these needs. It is here that the relationship with the level of the diffusion networks becomes clear. The members of the product development team need to know what the needs of the user of their product are: the user is seen as a source of innovation (Von Hippel, 1988) and interactions between the product developers and the users are needed. The product development team also needs to know about the technological options available at the technology developer. The supplier (in this study: the technology developer) is then a source of innovation.

2.3 The integrated model

An integration of dissemination and application implies an exchange between two levels: the level of the diffusion network for dissemination of a technology, and the level of the product development process for application of a technology. The integration of these two dimensions is visualised in Figure 1.

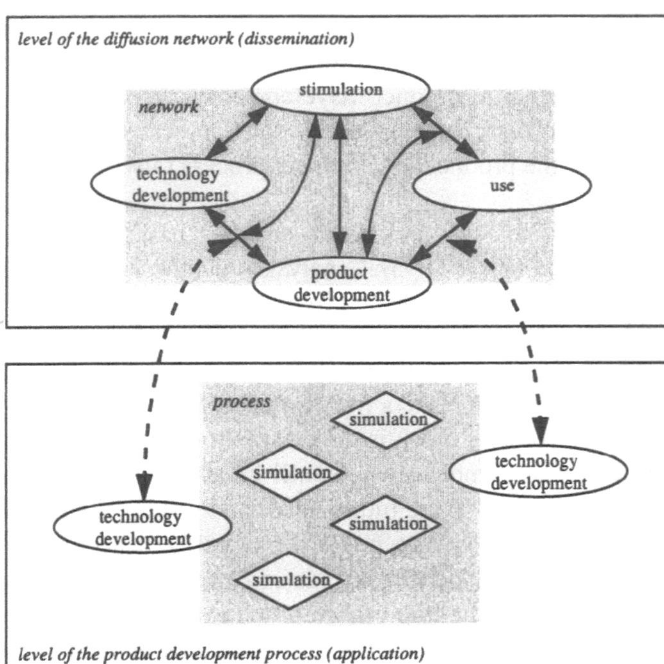

Figure 1. A theoretical model for analysing technology diffusion

The dotted black arrows show two examples of the connection between the two levels: the interaction between the technology developer and the product developer at the network level forms the external source 'technological options' for the level of the product development process, while the interaction between the product developer and the user forms the external source 'user wants'.

The integration of dissemination and application is possible when the actors active at both levels are able to understand each other, thus enabling knowledge transfer. The actors at the level of the product development process and their interactions within the diffusion network during the product development process can be analysed by describing the object worlds of these actors (Bucciarelli, 1988, 1994). Both the network interactions and the product development interactions are negotiations across the object worlds of the participants of these interactions. The participants represent a certain input into the negotiations. At the product development level, a designer represents the aesthetic values of the new product and the director represents the company's interests. At the network level, the participants represent their organisation with a specific function in the diffusion of a technology. Both technology dissemination at the network level and technology application at the product development level can be seen as social processes, where different actors participate with different backgrounds, different object worlds. The difference between the two social processes is that the participants in the product development process all have the same objective: the development of a common product. However, the participants at the network level might also have different or even opposite objectives in participating in the social process. To obtain a shared object world in such a network setting will require an extra effort.

The object worlds of the various actors active in the PV sector are different because of the different backgrounds of these actors. Two actors and their object worlds are shown in Table 1 on the left side. The object worlds, in grey, of these actors do not fit together. Both actors have to build an interface in their object world that might fit the interface of the other actor as can be seen in the middle of Table 1. When the interfaces of both actors are connected, as can be seen in the right column of Table 1, interactions can take place.

Table 1. Enabling and realising object world linking

two actors with different object worlds	interfaces are built by both actors (the dark grey parts)	interactions can take place when the interfaces are connected
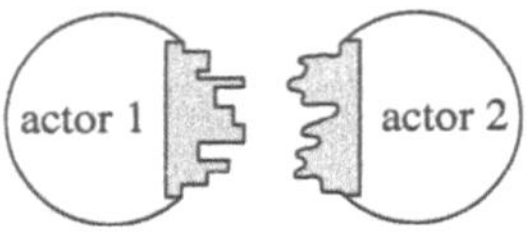	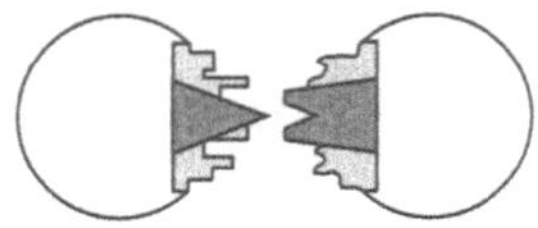	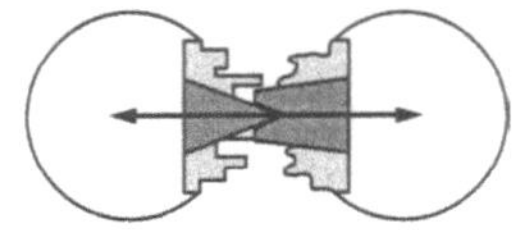

3. COLLABORATION IN THE DUTCH PV SECTOR

What is the actual situation of the PV diffusion in the Netherlands in the late 1990s? After a short introduction into the setting for PV in the Netherlands, the observations of the diffusion activities of the various actors in the Dutch diffusion network are described.

3.1 The Dutch setting for photovoltaics

The national research programme for photovoltaic technology was started in 1986. It centred on the development of photovoltaic technologies. Before, photovoltaics had only played a minor role in the first national research programme on solar energy of 1978. The most recent important year for photovoltaic technology was 1997, when the Dutch Ministry of Economic Affairs, responsible for the national energy policy, published its vision on sustainable energy in 'Renewable Energy – Advancing Power' (EZ, 1997). In this document, photovoltaic technology is viewed as the most important option for future electricity generation. The Netherlands Agency for Energy and the Environment (Novem), being one of the subsidisers for the Ministry of Economic Affairs, wrote a long-term plan for the development of photovoltaic technology in 1997 (Novem, 1997a). Government and industry also entered into the 'PV covenant' in 1997, in which they declared that they would make photovoltaic technologies competitive as an energy source for the 21st century (Novem, 1997b, 1998).

The generation of electricity by solar power in the Netherlands amounts to 4.4 MWh/year in 1997, but this is only 0.03% of the total Dutch electricity generation. The total amount of electricity generated by renewable energy sources is 0.3%. The use of biomass to produce electricity has the largest part in this: 80% (ECN, 1998). The yearly installed photovoltaic power doubled in 1996 compared to 1995; the largest increase was in the grid-connected systems (De Gooyer and Loois, 1997). The yearly installed photovoltaic power in 1997 was much the same as the installed power in 1996 (ECN, 1998). The Ministry of Economic Affairs, in their third Energy Policy document, defined a target of 10% saving of fossil energy through the use of renewable energy in 2020; the first target was set for the year 2000: 3% saved using renewable energy (EZ, 1996). Financial support was provided to realise these targets: an extra incentive was given to the 'Economy, Ecology and Technology' programme for the stimulation of sustainable technological development (EET, 1997), and 15 million guilders (6.8 million Euro) for a market introduction programme for photovoltaics.

3.2 PV diffusion in networks

The observations of the forty-five actors investigated in the Dutch photovoltaic sector are described below. The most important observations of the functioning of the actors are given according to the function of these actors in the diffusion network.

3.2.1 Technology development

Observations of the network function

The technology developers in the Netherlands can be divided into two groups, each with a specific function. The first group consists of universities and research centres; they have the function of cell technology development. These universities and research centres are focused on developing the technology of the photovoltaic cells. They invent new types of technologies and they explore the possibilities of using different materials. Triple junction solar cells and organic solar cells are examples of such new photovoltaic cells. The universities and research centres also test the technology, but only on a small scale.

The second group consists of manufacturers. These manufacturers focus on the production technology for producing the photovoltaic cells and combining these cells into solar panels. The producer of crystalline silicon cells was established more than 15 years ago and is showing an enormous growth, they want to have a 5% world market share by 2000. The second player in this field is still developing the production process for amorphous silicon cells, they will not start their production before 2005. The Netherlands is too small to have many big players in this field. Therefor, a third player in the manufacturing of photovoltaics has invested in foreign production capacity of amorphous silicon cells.

Most of the developments in both groups are financed by the national government. From a historical point of view, the main focus has always been on technology development. In general, the company size of the manufacturers can be considered large in comparison with the product developers, which are described below.

Observations of the network interactions

Most network interactions of the technology developers are concerned with interactions with other technology developers and stimulators, the subsidisers in particular. The product developer is more and more included in these collaborative settings. The user, the autonomous user in particular, is still absent. If there are contacts with users, these users are large-scale grid-connected users, such as energy companies.

The relationship with the product developers in the diffusion network concerns the application of the technology in photovoltaic systems. The largest manufacturer also performs the function of product development, they are a 'system house' and offer complete photovoltaic systems.

The relationship with the stimulators in the network differs with the 'age' of the organisation. Older technology developers have tentacles throughout the network with governmental bodies responsible for energy policy, with lobbying organisations and with environmental groups. Most of the technology developers are a member of one or more national or international solar energy associations. The branch organisation, however, is missing in their interactions. Some of them were a member but cancelled their membership.

3.2.2 Product development

Observations of the network function

The Dutch product developers are those actors in the diffusion network who apply the photovoltaic technologies. They compose the photovoltaic systems using the different components they obtain from Dutch or foreign suppliers. They design the system according to the needs of the user, produce and assemble the systems and supply them to their customers.

The development of photovoltaic product systems was in the beginning motivated by a need for independent energy supply and the fact that there were no such products available. In a later stage, it was also encouraged by a subsidy from the Dutch government, as in the case of the drinking installations for cattle. Other product developers started with photovoltaic products because they wanted (and were enabled because of subsidies) to explore the possibilities of large-scale applications, such as in Building Integrated PhotoVoltaics (BIPV). Other product developers made no use of subsidies and developed products that were commercial from the launch. An example of these kinds of products is a system for navigational aids.

'Pleasing the customer' is the mission of most of the product developers. But they also want to promote photovoltaic use, sometimes in combination with energy-saving goals. 'Everybody wants photovoltaics, but it should look good' and therefore, the design of high quality products is of major importance. A chance for photovoltaics lies in the use in recreational settings: 'People are willing to pay more if they do it for fun or their hobby'. Four of the product developers emphasised the importance of photovoltaics as a clean, reliable, low-maintenance and silent source of electricity.

Organisations with a product development function are smaller than those active as technology developers (Kruijsen and Bouwmans, 1998). The smallest is made up of only one man, the largest consisted of about 50 people. The majority not only sells PV, but a varied mix of renewable energy systems.

The fields of application of the Dutch product developers show a wide variety: navigational aids, street lamps, safety products, cattle drinking installations, sailor's equipment, roof systems and grid-connected Building Integrated PhotoVoltaics (BIPV).

Observations of the network interactions

The network interactions among the Dutch product developers show that all functions should be involved to ensure successful market introductions of new products.

Contacts between product developers and other functions in the diffusion network are different for each actor. The one-man business really is a one-man organisation: he has his contacts but prefers to work alone. At the other end of the spectrum are the product developers who operate like a 'spider' in a web. They have contacts with the technology developers who supply the solar panels, they have contacts with their clients who are the users of the photovoltaic systems they design, and they have contacts with a variety of different stimulators like governmental bodies, and financing and lobbying organisations. Three interviewees emphasised that the contacts with universities and other knowledge centres should be improved to stimulate the transfer of up-to-date knowledge.

It became obvious that the role of the user is of great importance. In product development projects where a potential client was not involved, the product idea did not make it to market introduction. If the potential client was involved and the product developer had a clear picture of the need of this user, the product idea was successfully introduced to the market (Kruijsen and Bouwmans, 1997). Personal contact with clients is very important. These clients are both autonomous and grid-connected users in the Netherlands.

The question "what could other actors do?" resulted in a variety of answers. The role of the branch organisation was criticised many times: it should operate more independently, not only promote the use of thermal solar energy, but also the use of photovoltaic solar energy; it should inform installers about photovoltaics (it is easy to install them, but they need to know more); it should organise more workshops and an internet plaza to couple the developers and the users; it should inform small and medium sized enterprises (SMEs) on state of the art of photovoltaic technology because it is difficult for them to stay informed. The subsidisers should reconsider their criteria: instead of looking at 'the amount of saved energy', they should give support in accordance with the amount of 'green' energy used in the new product or with the innovativeness of the idea if only a small amount is used. But if an innovative function can be realised, it should make photovoltaics more affordable. More small and medium sized projects should be subsidised and the quality of the assessment of subsidy proposals should be improved. It is considered really odd that the competitor of a company who applies for a subsidy may be asked to assess the proposal. Furthermore, standardisation is needed, especially in the BIPV, and education could play a significant role in the acceptance of photovoltaics by future users. A photovoltaic school-kit is one idea that could be promoted in this area.

3.2.3 Use

Observations of the network function

The users in the Netherlands can be divided into two groups: grid connected and non-grid connected or autonomous users. Photovoltaics fit in with the user's business or lifestyle. The users demonstrate that photovoltaics are operational in a wide variety of applications.

The users consist of individuals, one-man businesses, municipalities and large-scale utilities. Their experience with photovoltaics is mostly limited. Most users spoke about their first experience or were still working on their first project. The utilities learn fast, however. Within a few years they will be experienced users. The small-scale users bought their system, which is still working, and some basic knowledge of electronics is enough to maintain the system.

The first autonomous waterworks on photovoltaics were installed in the early 1980s, the first grid-connected system in 1983. New products like the cattle drinking installations were introduced in the early 1990s, and in the mid-1990s larger grid-connected systems were built, followed by electrical boats, street lamps, telecom poles and sound barriers.

The motivation to use photovoltaics is different for each type of user. The individuals who have photovoltaics on their roofs or on their camper or sailing boat, chose the green electricity because it is clean, reliable and quiet. The professional autonomous users decided to use photovoltaic systems on the grounds of their reliability and low maintenance. The reasons for using photovoltaics were 'economy' and 'reliability', such as in the case of farmers who have stand-alone cattle drinking installations and municipalities with small sluices working stand-alone on photovoltaics. Nature parks want to alert their visitors to the need for nature conservation which is an issue connected to the use of renewable energy so that they want to have their own electricity need supplied by photovoltaic energy. There also was the case of a municipality looking for a 'natural' way of opening and closing a bridge over a natural waterway. Most of the autonomous systems were installed without subsidy. The professional grid-connected users decided to use photovoltaics mainly because of the green image, their motivation for using photovoltaics is 'ecology'. They need the financial aid of subsidies.

The motivation of users of photovoltaic systems mostly fits in with their personal lifestyle or company mission. They are concerned about the environment or they supply sustainable goods and services and therefore also want to have a sustainable energy source for their own business. Users, especially the large-scale users, wish to be an example and a trend setter. Photovoltaics can make this world a better world, 'start with yourself', and reduce the CO_2 emissions, is the opinion of many users.

Observations of the network interactions

The network interactions are different for the users distinguished. The network interactions of the small scale, autonomous users are limited. They only have contacts with the product developer of their system. The utilities, the large-scale grid-connected users, have more contacts throughout the network, in particular with the stimulators. The utilities maintain close contact with governmental bodies, both at municipal and national levels, to obtain subsidies or other financial assistance. They sometimes have direct contacts with the technology developers.

Many users emphasised the need for financial incentives to make photovoltaics more affordable. Some would like a subsidy on photovoltaic products and product systems, others would like to see an extra tax on fossil fuels.

3.2.4 Stimulation

Observations of the network function

The fourth and last network function is the stimulation function. The stimulators in the Netherlands are those actors in the diffusion network who support the activities of the other three functions in the network. Two groups can be distinguished: (1) conditioning organisations and (2) intermediary organisations.

The conditioning organisations are governmental and financing organisations and provide conditions. Different programs are set up to take away technical and non-technical barriers. In the past, most efforts were put in technology development, recently market introduction has become more important.

The second group, the intermediary organisations, pass information to other actors in the diffusion network and couple actors with different network functions. They have been set up to increase the general public's awareness of the need for renewable energy, particularly solar energy.

Many interviewees emphasised that a strict division should be made between organisations that couple actors, the intermediaries, and organisations that subsidise projects of these coupled actors, to avoid a clash of interests.

Stimulation of the diffusion of photovoltaics does not always have a solely positive effect. Sometimes it has a negative impact, for instance when legislation discourages certain developments.

The goal of all the efforts by the stimulators is to increase photovoltaic use. Some of them work from the perspective of national interest, some of them from the perspective of the interest of parties that cannot speak for themselves: the environment and future generations. Some of them have commercial starting points, the banks in particular. All are convinced of the great potential of photovoltaics and this encourages them to continue with their work. Some of them are planning to achieve a certain result next year and mainly focus on a short-term; others take more time and set themselves goals for the year 2020 or even later. There is no correlation between the type of actor and the time horizon of their focus.

The size of the organisations ranges from a small-scale education centre, via a medium-sized consultancy to large-scale banks and ministries. The number of years experience with the promotion of photovoltaics range from more than ten years, via five years to only recent experiences. The stimulators have an interest in photovoltaic diffusion because they have committed themselves to the promotion of sustainable development. They want an increase in the use of renewable energy. Two of them particularly devote themselves to the promotion of photovoltaic use.

The output of the Dutch stimulators is a wide diversity of stimulating products. Two types of stimulating products are given below: conditions and intermediary. Conditioning organisations deliver for instance projects to develop standards, set goals such as 10% renewable energy in 2020, provide green financing, green labels, and energy tax and regulations, and research programs such as the Economy, Ecology and Technology programme (EET), and the Dutch national research and development programme for photovoltaics (NOZ-pv). Intermediary organisations develop promotional material, give support in the setting up of photovoltaic communities to produce photovoltaic electricity locally, publish articles in magazines which play an important role in familiarising the public with photovoltaics and photovoltaic products, and set up of 'central information points' where people can ask questions, and 'matchmakers' to couple actors.

Observations of the network interactions

The Dutch stimulators have relationships with all other functions in the network, as can be seen in the examples given below. The interactions of the stimulators concern the implementation of the above-mentioned stimulating products.

The intermediaries stimulate the development and application of solar energy through the exchange of information. They organise workshops and training sessions to disseminate the knowledge on photovoltaics and possible applications, and they publish newsletters and fact sheets. The branch organisation should connect the different network functions. However, none of the interviewed actors were enthusiastic about the coupling activities of this organisation. The branch organisation played an important role during the introduction of solar heating systems, but many actors, both technology and product developers, complain about the role this organisation has played during the introduction of photovoltaic systems. Many different actors are working hard on the stimulation of photovoltaic technologies, but sometimes the work is done twice and therefore co-ordination is needed.

The Netherlands is a member of international partnerships such as the European Union. Dutch actors participate in many international programmes such as the Altener programmes managed by DG XVII, the European Commission's Directorate-General for energy, and the International Energy Agency PhotoVoltaic Power System programme (IEA PVPS). Both programmes started in 1993.

The conclusions of the network interactions of the stimulators concentrate on stimulating the interactions within the network, removing barriers and increasing the familiarity with photovoltaics. All types of photovoltaic technologies are stimulated. The autonomous user is absent in the contacts of the government, but included in the interactions of the intermediaries. The intermediaries are scattered over the network, co-ordination is considered useful.

4. STRATEGIES FOR DIFFUSION

The Dutch photovoltaic sector can be characterised as a representation of the 'polder model'. The 'polder model' refers to the way Dutch actors are interacting. The Netherlands are a very horizontal country and so are the interactions: hardly any hierarchy can be distinguished. All actors, including the government, are supposed to be on the same level of importance, and, in general, they are very much a consensus oriented.

The emphasis used to be on technology development. Market development was put on the agenda only recently. The observations of the diffusion mechanisms as described above showed that the user was under-represented in the network. All interviewees strongly emphasised the need for market development. They made many suggestions. These points of interest cannot be detached from one another. All actors in the diffusion network have a function and only a portfolio approach, a combination of these points of interest, will lead to an enlargement of the photovoltaic market. The following list is an overview of the options for photovoltaic diffusion in the Netherlands as given by the interviewees.

4.1 General suggestions

The following points of interest concern all functions in the diffusion network. They aim at linking the object worlds of the different actors that have a potential role in the network:

- The 'mantra' for all the potential actors of the diffusion network should be: "Think in solar energy!" It is considered to be of great importance to have a clear vision of where to go and the slogan of the Netherlands Agency for Energy and the Environment gives a lead: 'The century of the sun starts in 2000'.
- 'Operate in networks' is the second important point of interest for all actors. None of the actors involved is able to stimulate the diffusion of photovoltaic technologies on their own. Connecting actors is also incorporated as a separate point of interest, see the next point.
- Actors must know why it is in their interest to participate in a diffusion network. This will convince them to participate in that diffusion network. Learning from earlier experiences, it becomes clear that it is important to adjust to local customs and to join in with existing patterns of needs, interests and relations.
- Learning programmes should be sufficiently flexible so that innovative learners who wish to go faster than others can do so.

The Netherlands plays an important role in the world with respect to the development of photovoltaic technologies. However, they should learn from their experience with wind energy. In the Netherlands, the emphasis has been on technology development for too long. It should be placed firmly on the implementation of the technology during product development and on market development. Photovoltaics are being developed in the Netherlands as a result of the investments made by the Dutch government, but it will take an extra effort to keep photovoltaics in the Netherlands.

4.2 Stimuli for the market

Stimulators can play an important role in preparing the market. The following strategies are mentioned as incentives to warm up the market and to simplify the investment in photovoltaic solar energy. All the strategies are directed at the user. The interviewees suggested six measures that can be taken by the government to stimulate the use of photovoltaics in the built environment (some of them already exist):

- implement a green mortgage, for example, a 2% lower interest rate for 10 years when a certain energy performance norm has been reached;
- make it obligatory to use photovoltaics in new buildings;
- introduce net-metering to simplify the return of overcapacity by users of a photovoltaic power station on the roof of their house or building;
- municipal governments should take photovoltaics into account at an early stage when developing zoning plans;
- the national government should set an example: all their buildings should be electrified by photovoltaic power;
- people are more easily convinced if they can see for themselves that a new technology really works. Give the public specimens of successful photovoltaic products to increase the belief in photovoltaic energy. For example, when someone achieves a 10% reduction in energy use, they will get a free photovoltaic specimen, for instance a watch.

The above measures are fairly easy to carry out. It will be more difficult to implement measures that have an impact on the entire energy market. In all countries surrounding the Netherlands, electrical heating is quite common. If the government was to promote the use of electricity in preference to the use of gas, the use of photovoltaics would stand a good chance of being introduced on a large scale. The question is whether this will be accepted by the Dutch energy market.

Another delicate subject is the price of electricity: because of the strong position of the existing energy market, the price of electricity generated from fossil fuels is low compared to that of electricity generated from alternative sources such as wind and solar energy. Preventive cost pricing, taking into account the costs of removing the damaging effects of, for example, the emission of CO_2 and including those costs in the price of electricity, is a still utopian concept in many countries, including the Netherlands.

4.3 Connections between developers and users

The exchange of knowledge and experiences is crucial for photovoltaic diffusion. To connect the different functions, the actors performing these functions should be invited to attend workshops with a specific theme, for instance a particular field of application for photovoltaics. An intermediary organisation, in its role of a stimulator, could invite people from the photovoltaic industry and people who are working in this field of application, both product developers and users. The follow-up after such workshops is important: projects need to be defined to develop the product ideas that emerged during the workshop.

Intermediary organisations like innovation centres function as a broker to secure the exchange of knowledge. Such organisations have an overall view of the knowledge available at universities and research centres as well as the contacts within the industry enabling them to see where this knowledge might be useful. Students can play an important role in transferring knowledge from the knowledge centres to the industry.

4.4 Familiarity of product developers and users with photovoltaics

If you do not know about the possibility of using photovoltaic technologies, you will not think of using them. Here lies the importance of a well-functioning promotion organisation. A branch organisation should support the photovoltaic industry and provide product developers, who might be able to use photovoltaic technology, with information about the technology, and it could promote the use of photovoltaic solar energy in a broader sense. Other organisations, such as groups belonging to the environmental movement, could also inform the public with information about the use of photovoltaic solar energy, as long as they give objective and realistic information.

An intermediary organisation like a branch organisation should operate as an information centre serving small companies who are unable to keep up with all the new information about photovoltaic solar energy. This information centre could be something like an internet plaza for the whole photovoltaic sector. It is not only a passive information supply that is important, but the active supply of information, such as through publishing about photovoltaic solar energy, is also vital.

The general public as well as specific users, such as the building industry, need to be provided with information about photovoltaic solar energy. A good start can be made by educating children who often are more environment-conscious than their parents. Intermediary organisations can provide schools with a photovoltaic solar energy kit containing a photovoltaic panel and a few applications like a small radio or fan. Other potential users should be enabled to follow symposiums and training sessions. Special training material should be developed for the education of installers, and the internet plaza and publications mentioned above should also have an informative function.

4.5 Financial incentives

The Dutch government is aiming at an increase in the use of photovoltaics by a factor of 10 between 2000 and 2020. Financial incentives could play a major role in achieving this goal provided they are designed and implemented carefully.

The subsidies should be given based on the innovative nature of the result of the project. Absolute figures like the amount of energy saved or the amount of green energy that will be produced, may restrain developments that could lead to new types of applications of photovoltaic solar energy.

Subsidies should also focus mostly on the user: make photovoltaics easy and affordable. These subsidies should decrease in due course to stimulate an independent market after the subsidy has been taken away.

4.6 Product development

A photovoltaic design tool needs to be developed that can be used to guide developers while they are developing new products that use photovoltaic solar energy. This tool should include information about square meters of photovoltaic panels required, cost-prices, energy pay-back time, etc. It should enable developers to implement the input they get from the users (the wants) and from technology developers (the technological solutions) into their new products or product systems.

A product developer may take the following items into consideration when designing photovoltaic systems:

- people want photovoltaics because it is visible;
- people are willing to pay more for a product or product system if they use this product for fun or for their hobby;
- people are willing to pay more for quality and the use of photovoltaics is easily recognised as high-tech and high-quality;
- if people are unable to invest in a complete photovoltaic system, they should be able to invest in a part of the system to begin with and add the other part(s) later. For example, it is possible to buy the tiles only and add the photovoltaic panels later;

- the reduction of the power needed for a product is a big incentive to use photovoltaic solar energy;
- people buy photovoltaic products firstly because of the price, secondly because of the reliability and only thirdly because of the clean energy;
- always work together closely with your clients because they know exactly what they want. Close interaction with the supplier of photovoltaic components will guarantee an optimal design.

5. CONCLUSION: TOWARDS A BREAKTHROUGH IN THE USE OF PHOTOVOLTAICS

The central conclusion of the exploration of the Dutch photovoltaic sector is that collaboration in the Dutch PV sector is a precondition for effective diffusion. The user was not always sufficiently involved. Market development is the missing factor in photovoltaic diffusion. But how can collaboration in the PV diffusion network be obtained? In the previous section, five strategies with regard to the stimulation of market development have been formulated; a sixth point of interest, the financial incentives, is a prerequisite for realising these five points.

The *overall* strategy to obtain collaboration is linking the object worlds of the actors in the diffusion network. All four identified functions in the network have to be present in the diffusion network, including use, because none of the actors performing one or more functions are able to steer the diffusion of photovoltaic technologies on their own. All the actors should think in solar energy and operate in networks to expand the Dutch photovoltaic sector.

REFERENCES

Bucciarelli, L.L. (1988) An ethnographic perspective on engineering design. *Design Studies* 9-3:159-168.

Bucciarelli, L.L. (1994) *Designing Engineers*. Cambridge, MA: MIT Press.

De Gooyer, H. and G. Loois (1997) *Marktoverzicht zonnestroom in Nederland*. Utrecht: Ecofys.

ECN (Energy Center, the Netherlands) (1998) *Energie Verslag Nederland 1997*. Petten: ECN.

EET (Economy, Ecology and Technology programme) (1997) *Economie, Ecologie en Technologie*. Utrecht: Programmabureau EET.

EZ (Ministry of Economic Affairs) (1996) *Derde energienota*. Den Haag: Sdu Uitgeverij.

EZ (Ministry of Economic Affairs) (1997) *Renewable Energy – Advancing Power*. Den Haag: Economische Zaken.

Håkansson, H. (1985) *Industrial Technological Development; a network approach*. London: Routledge.

Kruijsen, J.H.J. (1999) *Photovoltaic Technology Diffusion; Contact & Interact.* Delft: Eburon (dissertation).
Kruijsen, J.H.J. and I. Bouwmans (1997) *Let the sun shine! Introduction of solar cell technology into the product innovation process*; Paper for the 1997 International Sustainable Development Research Conference. Manchester, April.
Kruijsen, J.H.J. and I. Bouwmans (1998) *Strategies for the Diffusion of PhotoVoltaic Technologies* Paper for the 2nd World Conference and Exhibition on Photovoltaic Solar Energy Conversion Vienna, July.
Kruijsen, J.H.J., I. Bouwmans and M.P.C. Weijnen (1998) *Sunny Developments, the Diffusion of PhotoVoltaic Technologies in the Netherlands.* CD ROM with the proceedings of the 7th International Conference of the Greening of Industry Network; Rome, November.
Lundgren, A. (1995) *Technological Innovation and Network Evolution.* London: Routledge.
Novem (1997a) *Samen op weg naar de eeuw van de zon; NOZ-pv 1997-2000.* Utrecht: Novem.
Novem (1997b) *PV-convenant.* Utrecht: Novem.
Novem (1998) *Eén jaar PV convenant.* Utrecht: Novem.
Roozenburg, N.F.M. and J. Eekels (1995) *Product Design: Fundamentals and Methods.* Chichester: John Wiley & Sons.
Von Hippel, E. (1988) *The sources of innovation.* Oxford: Oxford University Press.

Chapter 10

A MODEL OF AN ENVIRONMENTAL BUSINESS SUPPORT NETWORK

Evaluating a Partnership Approach in Industrial South Wales

FRANCES HINES

1. INTRODUCTION

In the United Kingdom, the 1990s witnessed a proliferation of government funded agencies, non-governmental organisations, consultants, environmental pressure groups and individuals providing support for business to improve its environmental performance. The most common objective of these organisations was to raise awareness of environmental issues within business, and to facilitate the integration of environmental concerns into the management process. The increasing number and diversity of organisations providing such environmental business support has led to a bewildering situation for company managers, who are often confused about the type of support available.

This chapter presents a model of environmental business support as developed in South Wales. The model was developed as an attempt to provide an alternative to the existing poorly co-ordinated support available in Wales. Partly in response to the confusion generated by the number of environmental business support services, interest was expressed by a group of organisations in collaborating to achieve a more coherent approach to environmental business support services. In addition, it was hoped that a co-ordinated programme of strategic planning between the partners would

T.J.N.M. de Bruijn and A. Tukker (eds.), Partnership and Leadership, 177–193.

enable more effective targeting of funding than had previously been achieved.

In this chapter the motivation and goals for the new partnership approach are described. The ways the model worked in practice and what effects it had are explained. Following this, the strengths and weaknesses of the model are analysed.

2. WHY WAS A NEW PARTNERSHIP APPROACH NEEDED?

A number of weaknesses characterised environmental business support in Wales during the mid to late 1990s, as they did elsewhere in the UK. Duplication of services and projects occurred, often with very similar objectives. As a result, some parts of Wales suffered from a lack of support whilst others appeared over-provided. The plethora of support organisations meant that even those people at the top of the environmental policy and administrative hierarchy in Wales were not always fully aware of all of the organisations, programmes and projects aimed at improving the environmental performance of business. Among the support organisations themselves, there was a lack of awareness about the existence of other organisations, and the support they provided for business. In addition, support organisations lacked knowledge of the nature and detail of proposals for funding submitted by other agencies; and as a result a degree of conflict had arisen between many of these different organisations. This situation was recognised by a number of academics, administrators and practitioners in the field working within the Welsh Development Agency WDA and the former Welsh Office.

Baylis, Connell and Flynn (1997:22), for example, in their large-scale analysis of data gathered from SMEs and large companies in industrial South Wales in 1996, showed that companies identified a wide range of organisations providing environmental information and advice. However, Baylis et al. argued that environmental business support organisations did not perform well in the view of Welsh SMEs. Only one organisation, the Wales Environment Centre (in tenth place), appeared in the top eleven organisations identified from a total of twenty-two named sources of advice, whilst other environmental business support organisations were ranked substantially lower. SMEs identified waste management companies, local authority Environmental Health Officers, and books and journals as being far more useful. The reasons for such an apparent lack of knowledge about environmental support agencies centred around the confusion created by the large number of organisations in operation, and the weak referral system from economic business support agencies. Baylis, Connell and Flynn suggested that their work confirmed the findings of Christie et al. (1995) in the latter's assertion that "external advice and information services may not be seen as useful or approachable, especially by SMEs short of time and finance for seeking advice" (Christie et al., 1995:78 cited in Baylis, Connell and Flynn 1997:23).

As a result of the apparent confusion and unequal distribution and application of environmental business support services within Wales, a number of key organisations decided to make an attempt to develop a model of co-ordination and partnership that could address these issues. A decision was reached to bring together "for the first time in the UK, a partnership between the academic expertise of colleges and universities, business support groups and practical environmental management organisations to help small businesses [in South Wales] to improve their environmental awareness and performance" (Environmental Services Partnership, 1997a:2). This move towards increased partnership between support agencies was seen as a logical step. As Martin and Oztel (1996:132) argue, a local co-ordinated approach is thought to be more effective than a traditional hierarchical approach, which existed in Wales especially because of the nature of European funding. Co-ordination would provide a "visible focus for small business support" within a local area (Moran, 1993:131). In particular, it was thought that promoting co-operation between support organisations would produce a leaner, fitter and more accessible co-ordinated environmental business support network for Welsh companies.

3. HOW DID THE PARTNERSHIP WORK?

The Environmental Services Partnership (ESP) was designed to work on two levels. First, a partnership within a Steering Group structure between some of the main environmental support providers in Wales was set up, to more effectively target funding for research and environmental business support programmes. Second, as an extension of the above partnership, a partnered project was created to provide focused environmental support for small and medium sized enterprises in South Wales delivered by a co-ordinated network of some of the partners. The model was initiated in 1996 and the project component was due to end in December 1998, although it was expected that the model, once established, would form the basis for environmental business support partnership in Wales in the future.

The partners involved in the Steering Group were drawn from each of the three levels shown in Figure 1, which provides a simplified structure of environmental business support in Wales.

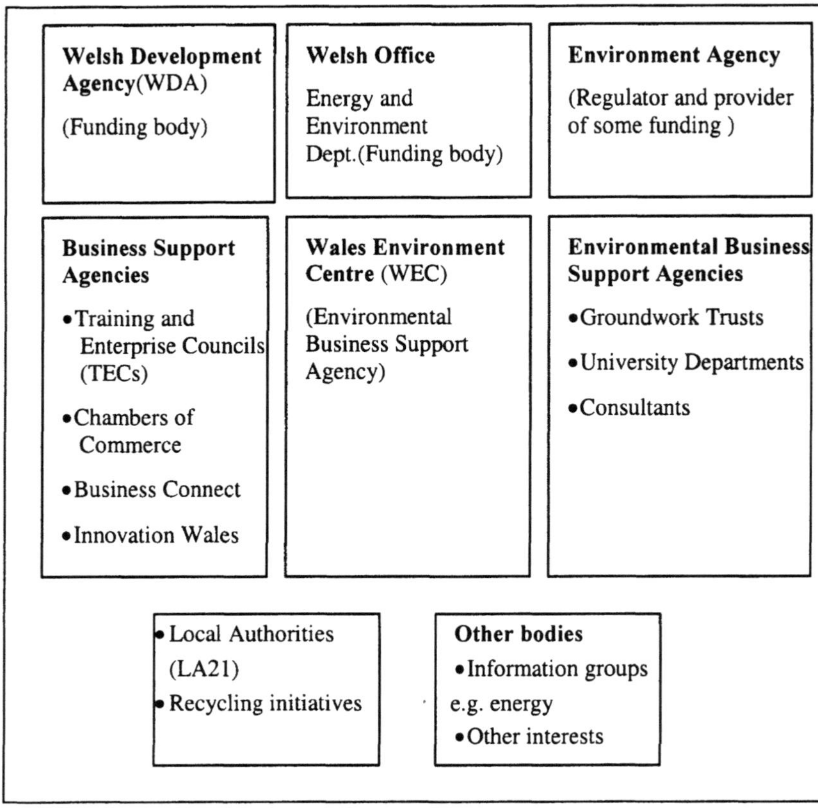

Figure 1. Simplified structure of environmental business support in Wales

At the top level is the National Assembly for Wales (previously known as the Welsh Office) and the Welsh Development Agency (WDA). These policy bodies have a degree of influence over the direction of environmental business support in the Wales, not least because they are often the channels through which funding is delivered to other agencies. The Environment Agency Wales influences the direction of business support through its enforcement of regulation, but also in its role as a funding body, and declared its intention in the late 1990s to become more directly involved in a business support role. At the second level, there are the main business environment support organisations. Arena Network (formerly known as the Wales Environment Centre (WEC)), and Groundwork Wales are the most important. A large number of projects managed by the Universities are included in this group. At the third level lie the Local Authorities. This group of support providers tend to exist on the periphery of the support network. They are generally poorly funded and few Welsh Local Authorities have had sufficient resources until recently to develop extensive support programmes for local businesses. There has been some co-operation between the main environment business support agencies, such as the WEC and some Local Authorities, but it has always been on a small scale.

4. THE ENVIRONMENTAL SERVICES PARTNERSHIP MODEL: CASE STUDY

The ESP Steering Group (as shown in Figure 2) aimed to draw together a range of organisations which regularly sought funding from the European Regional Development Fund (ERDF). These organisations included university departmental research groups from two universities, a local tertiary college, business and environmental non-profit support organisations, including Innovation Wales, Groundwork and the Wales Environment Centre, an environmental consultant and the Welsh Development Agency. Prior to the ESP Steering Group, funding proposals were developed individually by each organisation with little or no reference to the applications being submitted by others. As a result, there was often duplication of ideas and some similarity between proposals, leading at times to a conflict of interest between different agencies. Thus, the ESP Steering Group was set up as a "forum consisting of organisations in receipt of European funds that were specifically targeted at improving the environmental awareness and performance of SMEs is South Wales" (ESP Information Sheet, 1997). It aimed to reduce the duplication and the conflict of interest that had arisen before the partnership. The partners were to meet formally every two months to discuss project activity and related issues, especially in relation to submission of applications for ERDF funding.

The second level of the ESP model, the partnered project, was designed to provide free environmental business support to a large number of SMEs in South Wales. The project received funding from the ERDF. One of the main objectives of the project was to provide SMEs with a single point of contact through which they could access a number of the main environmental business support organisations existing in Wales thus reducing duplication. The Director of Innovation Wales, the main administrator of the ESP project, suggested that the model was distinct from environmental business support networks that existed elsewhere in the UK because:

> "each of our partners are experts in their own right but until now, each partner presented similar services under separate flags...but now we can cut out the duplication and present a coherent service to small businesses across South Wales. We all want to help SMEs to improve their environmental awareness and performance, together as a Partnership, our message is clearer and we can do a much better job." (Environmental Services Partnership, 1997)

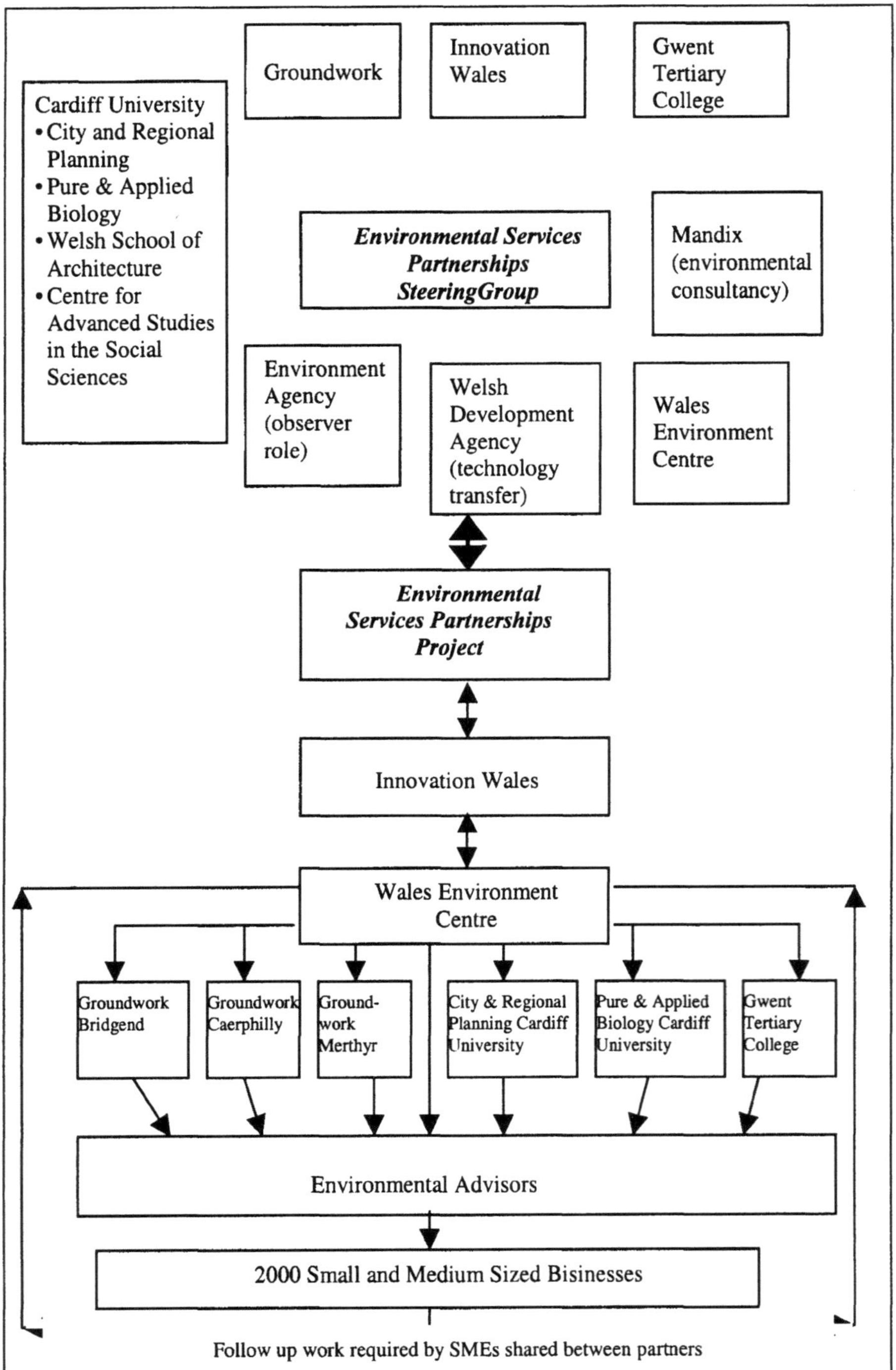

Figure 2. Structure of the ESP model

The partnered project linked directly to the ESP Steering Group, and issues that arose were expected to be dealt with at both of these levels in an open and positive manner. The co-ordinated network of service providers generated by the ESP partnered project offered a comprehensive range of supporting services to address all environmental aspects including training, information, advice, contacts, pilot programmes and financial assistance. The ESP project had a range of objectives and targets that it had to achieve by the end of 1998. The objectives were to:

- raise the awareness of environmental issues throughout the SME business community;
- assist businesses to manage their activities in a more environmentally responsible manner;
- encourage businesses to manage their resources more sustainable and;
- assist businesses to be more competitive by cutting costs and improving their market image.

The project targets included: visiting 2,000 SMEs by the end of 1998 to provide assistance on environmental legislation, technology and supporting services; establishing network groups to facilitate best practice; establishing two supply chain projects to bring suppliers and customers together; and supporting 2,000 SMEs in improving product and process efficiencies, innovation improvements and marketing assistance where feasible. The project was delivered by environmental advisers attached to the Steering Group partner organisations. Advisers visited targeted businesses, carried out a simple environmental site audit to identify specific problems facing the company, and followed up the visit with a written report detailing actions that the company could take to improve their environmental performance. This report would, where appropriate, name the partner organisations of the ESP project as service providers for particular problems. Information was contained in a database accessible to each partner organisation. This level of co-operation was unparalleled in previous ERDF funded projects in Wales.

The project was delivered through Innovation Wales, a non-profit organisation which was originally set up to stimulate innovation amongst companies in Wales. It was chosen as the project deliverer because the original proposal for funding had come from people working within the organisation. The choice of Innovation Wales was not greeted with universal acclaim. In fact, the Welsh Office was quite clear about its wishes in the matter: it wanted the Welsh Office funded one stop shop Business Connect network to deliver the project. Business Connect would seem to have been a logical choice of service deliverer, but in fact, due to lack of knowledge and experience in environmental issues, it could have had little more than an initial signposting role to a small number of environmental business support organisations.

Despite the wishes of the Welsh Office, the project was allowed to stay with Innovation Wales, who contracted the administration of the project to the Wales Environment Centre. Seven of the ESP Steering Group partners were invited to provide a representative of their organisation to work as an Environmental Adviser to deliver the project to the SME community. The number of advisers fluctuated slightly during the life of the project achieving a maximum of eleven people. Figure 3 shows how the partnered project worked, and how the environmental advisers ensured the different partner organisations were tied into the information provided to each business.

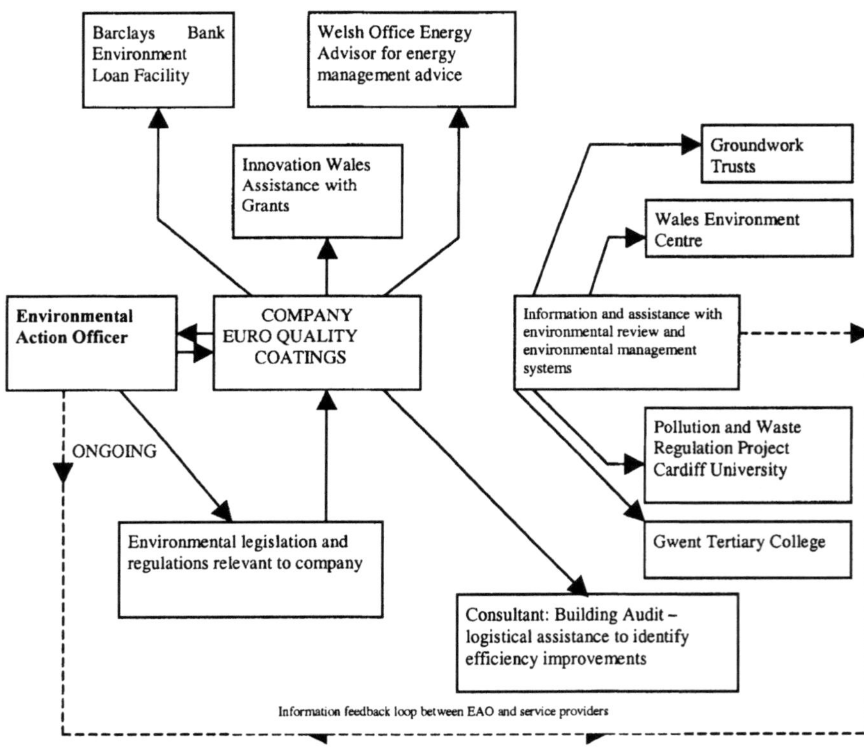

Figure 3. Connections for SMEs to the environmental business support network through the Environmental Services Partnership Project

5. BENEFITS OF THE ESP MODEL

The ESP Steering Group showed early indications of beneficial outcomes. Meetings during 1996 and 1997 indicated a clear commitment by the partner organisations to the idea of co-ordination of funding proposals. At these meetings, organisations provided information about proposals they wished to submit and were open to discussion when potential conflicts with the work of other agencies arose. However, the beneficial outcomes were much more tangible from the partnered project. Over 1500 SMEs were visited by the environmental advisers working for the project. Whilst much of the support was of an advisory nature, a significant number of companies acted on the advice and developed projects or strategies to directly improve their environmental performance. These included carrying out environmental reviews with support assistance, increased interest in environmental management systems, environmental policies and increased attention to compliance with environmental regulation.

A further beneficial outcome of the partnered project was the development of links between the environmental advisers representing the member organisations. Through group training meetings and monthly update meetings, this group of people made an effort to maintain communications. One result of this was a strong working relationship that encouraged the pooling of information and the effective referral of companies and company problems between organisations. The Environmental Advisers commented that they "have gained much more knowledge about other support agencies" and that "awareness amongst SMEs about agencies and about environmental issues has been good" (comments from interviews with environmental advisers in September 1998).

6. WEAKNESSES OF THE ESP MODEL

The main area of weakness appeared to lie with the ESP Steering Group level of the model. The different organisations, which were ostensibly committed to discussing their funding proposals openly with the other members of the group, did not find it easy to put this into action. Whilst intentions may have been good at the outset, the latter stages of the implementation of the model witnessed a significant withdrawing of interest in continuing the exercise, and a return to the less open funding application situation that existed before the advent of the ESP. A comment by one of the representatives attending the ESP Group meetings was that "[the ESP Group] has not broken down barriers...[and people] find it difficult to lay aside their own agendas". A clear illustration of this lessening of commitment was the infrequency of ESP Group meetings held in 1998. Following a meeting of the ESP Steering Group in September 1997, its members, whilst individually agreeing that the co-ordination of business support services was necessary, exhibited a marked disinclination to continue sharing information regarding their submission proposals for funding.

Causes of dissension and the partial return to the situation that existed previous to the ESP were clearly related to personality and the desire to retain status and power within the support network. One of the main organisations involved in the ESP project was the Wales Environment Centre. The WEC is generally recognised in Wales as one of the central organisations involved in providing business environment support, and has itself been involved in a partnership with a number of other organisations of a governmental (national and local), commercial and business support nature (Business in Wales, 1998:24). It had not, up until the development of the ESP, been involved in a formal partnership with other environment business support providers. The WEC, along with other organisations involved in the project, indicated on more than one occasion that difficulties arose in working with people providing similar services especially when the perception of the quality of work carried out by these other organisations was not very positive. The Steering Group and the partnered project activities of the ESP may have enlightened members about the services offered by other organisations, but, as one organisation employee stated, "it has not made me esteem them any higher".

Funding was another cause of dissension. The ESP partnered project was funded through the ERDF and required matched funding. This was a continuing problem for all of the organisations involved. It may have been instrumental in the dissolution of the contract between Innovation Wales and the WEC that occurred in March 1998, as the latter organisation felt it could not provide further matched funding and thus declined to continue in its role as project administrator. The picture may have been more complex however. One of the representatives on the ESP Steering Group felt that the WEC had, in fact, used the funding issue as an excuse to pull out of the contract because it was unhappy about the position it placed the WEC in with the Welsh Office. As the Welsh Office had never been completely happy with the delivery of the project through an organisation other than Business Connect, the WEC may have felt that their continued association with the ESP project put them in an awkward position with their main supporter. Other members of the Steering Group felt that there was a further danger associated with the funding of organisations that were essentially going to carry out activities that could have equally been undertaken by commercial consultancies. The question was asked as to whether public money should be spent on services that should be provided by consultants (comment from telephone interview with ESP Steering Group representative).

An illustration of the funding problem is provided by discussions surrounding the idea of a constitution for the ESP. There was "extensive discussion" in 1996 to determine whether a formal constitution was needed. Discussions within the ESP Steering Group centred around the argument that a formal constitution would "assist in obtaining matched funding and [would] provide a focus for services when ERDF ended" (Environmental Services Partnership, 1997b). The issue of whether the ESP should have a constitution was never resolved, and was a pivotal weakness as it meant that the continuation of the partnership beyond the European funding of the ESP partnered project to the end of 1998 was never assured.

The partnered project also suffered from a range of problems. The lack of clear communication, and latterly commitment, between the ESP Steering Group and the environmental advisers working on the project was disappointing. One environmental adviser felt that he had "little awareness of what the [Steering] Group does", whilst another adviser commented that the Steering Group "could have been more supportive and involved...[and that] a greater flow of information could have been directed to the ESP Project" (comments from interviews with environmental advisers in September 1998). There appeared to be a general feeling that the problems besetting the Steering Group had a negative influence on the project, both in terms of support and communication, and in terms of leaving those people working who were on the project, unclear about the relationship between the organisations involved.

There were also problems relating to the administration of the ESP Project. The Steering Group did not appear to have given sufficient consideration to the logistical requirements of implementing such a large scale partnered project. One environmental adviser felt that the Steering Group "underestimated the level of co-ordination that was going to be required". As a result, the provision of company information for contact purposes and the development of an accessible Web based database for member agencies suffered from continual problems.

Thus, while there were clear benefits from the implementation of the model, there were also substantial weaknesses which, whilst not affecting the service provided to business, could have had negative impacts on the long-term relationship between participating organisations. The model appears to have both succeeded, and failed, on a number of levels. Where the partnered project succeeded was in providing environmental support to a very large number of companies in South Wales. At the end of 1998, such a widescale impact on the SME community had not then been equalled in the UK.

Co-operation did develop between the environmental advisers delivering the partnered project, and these relationships continued beyond the life of the project. Organisations undoubtedly became more aware of the support services being offered to business by partner organisations and were increasingly sensitive to the need to share ideas and resources to achieve efficiency in delivering assistance to business.

7. ANALYSIS OF THE MODEL

The rather ad-hoc way in which support has developed in the UK, the variability in spatial provision, the diversity in specialisms offered by different organisations and the difference in quality of support provided, (even within one support network), has resulted in a bewildering provision of support services that the average business manager finds it difficult to use to his best advantage. Chubb (1998:11), whilst describing business environment networks (groups of businesses organised into clubs for mutual support), identifies some of the essential problems characterising such networks when he points out that in many areas where funding is plentiful, European Regional Development Fund areas in particular, "a profusion of [Business Environment Networks] have established themselves, often with overlapping boundaries". He further argues that this may create "an atmosphere of intense competition" between the different agencies, and that support becomes involved in a "numbers game", with SMEs developing an image of aggressive competition between support organisations. This can often result in duplication of contact to busy SME managers who rapidly become disillusioned with the lack of clarity exhibited by the environmental business support system, and who then may take little advantage of the benefits it does offer. Hutchinson and Chatson (1994:20) suggest that the profusion of support organisations has meant that "support is probably the last thing they achieve in supplying". One of the main reasons for the confusion generated is that mechanisms for referral between organisations have only recently begun to be developed and are still weak in many localities. In addition, where there are no formal means of integration between organisations, it is often up to the individual organisation to formulate priorities and define areas of responsibility (Segal et al., 1988:47). Many organisations that are reliant on funding bodies for their continued existence may feel it necessary to demonstrate the quantity of work they achieve to those funding bodies. This may mean they are reluctant to pass work on to other organisations, even if those organisations have greater expertise or specialist knowledge.

Development of a network in which organisations co-operate and co-ordinate to produce an integrated service which will provide the most effective and efficient support for business is dependent on trust, good relationships between key personnel in each organisation, a recognition of common aims and transparency of action and motive in relation to funding proposals for projects. The difficulty in meeting all of these criteria means that the situation in most localities is still one of defensiveness between agencies trying to ensure their existence.

Such defensiveness was ultimately one of the reasons why the ESP model failed in its aim to establish a more structured approach to the provision of environmental services to business, but especially a co-ordinated approach to funding. The tangible successes of the partnered project were overshadowed by the obvious failures of the implementation of the overall ESP model. The concept on which the model was built was sound, but did not take sufficient account of the strong personalities of organisation managers who struggled to reconcile their desire for co-operation with their need to retain organisation status. The model was developed without sufficient attention to the differing agendas of the participating organisations, and limited consideration of the existing relationships (some of them of considerable duration) between the managers and leaders of the organisations involved in the model. Organisations were, and remain, under constant pressure to respond to the 'quantity not quality' criteria that characterises funding from European sources (notably ERDF) and it is often this pressure that motivates and controls the strength and dynamism of these relationships.

In an increasingly saturated market, there is no real motivation for organisations to voluntarily place themselves in a position which they believe may erode their stature in the eyes of the funding bodies. Instead, this climate of competition appears calculated to make organisations more territorial about their areas of responsibility, and more defensive in situations which require them to release that responsibility to other organisations. Whilst it is generally understood that a co-ordinated approach to funding would, in theory, provide a more focused and potentially efficient service for business, it is very difficult for organisations to expose themselves to any action that may reduce their influence or status within the overall environmental business support network in the country.

The degree to which businesses are actively involved in the development and implementation of environmental business support, either at the level of a steering group, or in the development of a project, is an important factor in determining the success of co-ordinated ventures. The ESP model singularly failed to include an industrial partner at any stage or at any level, instead treating business as a generally passive recipient of advice and support. Subsequent projects carried out by the different partner organisations have revealed how crucial it is to consult businesses from the very beginning. While such an attitude may seem obvious, the reality of the situation in Wales was, and to some extent remains, one in which support organisations see themselves as being somewhat apart from business in general, despite the need to provide a co-ordinated and responsive approach. They are very much in the business of providing supply-side driven support. The demand-side is rarely considered and is the source of many assumptions and little objective research.

The ESP model suffered from too many fundamental weaknesses to allow it to achieve its original aims. While it had the potential to be a transformational model, especially in the way it attempted to provide a co-ordinated structure and approach on two levels, problems were generated by a poor understanding of the dominance that could be exerted by key figures with strong personalities. In addition, the limited communications between Steering Group and partnered project personnel, the lack of ownership of the model felt by many people working within the partnered project, the poor appreciation of the dynamics of power as influenced by external agencies, and the lack of business involvement in the delivery of the project ensured that it would have been very difficult for the model to succeed.

8. CONCLUSIONS

As environmental support for business at a local or regional level is a recent phenomenon in the UK, relatively little research has been carried out in identifying the patterns and problems characterising environmental support as distinct from those identified in the provision of economic support. These have obviously rather different objectives, but there are similarities between them and useful parallels may be drawn. One such similarity is the lack of co-ordination in the growth of the support networks to be found within a region. Deakins (1996:137) describes the development of the enterprise agency movement, for example, as "ad hoc" and suggests that "the provision of support, spatially and vertically, is the result of chance and accidents of geography". A number of problems have arisen as a result of this uncoordinated growth, one of which has been that a number of support services provided by a range of organisations can develop within a region, but that they often operate at a very local level. These organisations may have little knowledge about others operating within their region, and may also have a low level of awareness about the range of business support services and projects available to the businesses that they serve.

The Environmental Services Partnership model implemented in Wales was a brave attempt to address some of the issues that characterise environmental and economic business support networks in the UK in recent years. It set out to try to co-ordinate action on a number of levels: notably by establishing a forum in which potential conflicts in funding applications could be discussed and hopefully resolved, but also by providing a business support network which provided a single point of contact for businesses, offering easy access to a range of support organisations.

Although the criticisms levelled at the implementation of the model appear to outweigh the apparent benefits it has produced, the potential for benefits to accrue do exist. To assure success, however, certain factors would need to be considered more fully than they appear to have been in the Welsh case. The issues surrounding organisational responsibility and control of resources need to be brought to the attention of potential partners before the initiation of the model. Clear groundrules need to be put in place to ensure that no one group derives more or less benefit from the actions taken than any other group, and emphasis must be placed on the need for greater transparency and trust in organisational relationships. Logistical and administrative demands placed on organisation members need to be recognised, and communication between organisations needs to be developed in a structured way to encourage continued and effective communication thus avoiding the breakdown of relationships.

The principles which underlie the development of the model are sound. It is recognised by many organisation managers, advisers and other network actors that co-ordination and co-operation are inevitable if greater efficiency and effectiveness in supporting businesses to improve environmental performance is to be achieved. Whether this is through the medium of the one stop shop or through closer relationships between a range of national and local organisations depends on the current structure and form of networks within a region. In some European countries, the business support networks that already exist are much more structured and coherent than those found in the UK. If such a model were to be implemented in another region of the UK, the weaknesses that have characterised its use in Wales would need to be fully addressed if benefits were to be gained. The main problems that face the implementation of such a model are that in many cases, whether in the UK or elsewhere, substantial changes in the way in which business support organisations are funded are necessary. The current ethos of funding dependent on quantitative results does not encourage the development of quality services. If organisations are constantly under pressure to produce results 'by numbers', they have little motivation to co-ordinate their efforts with those of other organisations: the market becomes too competitive as business support providers struggle to increase their client group amongst a limited number of businesses. It is also difficult to effect necessary change in the management culture of organisations. Whilst many operate through a non-profit status, managers are often people with commercial management experience and they may therefore find it difficult to divest themselves of the competitive nature inherent in business management, as co-operation between businesses to deliver products or services is not a common occurrence. As a result, administering a large scale project between a number of organisations requires a culture of communication and co-operation that, in the UK at least, appears to be very weakly developed.

While the ESP model produced some benefits, mostly short-term direct benefits to business, it did not achieve its long-term aims of co-operation and co-ordination of environmental business support in Wales. The model, whilst highlighting many of the weaknesses of the existing business support network, did not change the fundamental structure or nature of relationships between participating organisations. As an attempt to evoke transformational processes within business, it undoubtedly had some success, although the impacts may have been rather superficial in the majority of cases. Even in this process, it cannot be said that it was the effectiveness of the partnership that stimulated changes in business. The changes were more a result of the scale of the partnered project and had little to do with the Steering Group or the overall aims of the model to achieve a co-ordinated support network. Indeed, many of the flaws of the environmental business support network in Wales may have been further reinforced since the demise of the ESP model by the next tranche of ERDF Objective 1 European funding, although a co-ordinated attempt between two of the original ESP partners and some of the Local Authorities in Wales looks set to be a more successful venture.

REFERENCES

Baylis, Robert, Lianne Connell and Andrew Flynn (1997) Environmental Regulation and Management: A preliminary analysis of a survey of manufacturing and processing companies in Industrial South Wales *Papers in Environmental Research Department of City and Regional Planning*, Cardiff: Cardiff University.

Business in Wales, *Western Mail* 1998; 10:7:24.

Chubb, S. (1998) Business Environment Networks for SMEs *Local Environment News* 4, 1: 9-11

Christie, Ian, Heather Rolfe and Robin Legard (1995) *Cleaner production in industry: integrating business goals and environmental management.* London: Policy Studies Institute.

Deakins, David (1996) *Entrepreneurship and small firms*. Maidenhead: McGraw Hill Publishing Company.

Environmental Services Partnership (1997) *Information Sheet for the Launch of the Project on 10th August 1997*. Cardiff: ESP.

Environmental Services Partnership (1997a) *Press Release - Environmental First for South Wales 10th September 1997*. Cardiff: ESP.

Environmental Services Partnership (1997b) *Minutes of ESP Group Quarterly Meeting September 5th 1997*. Cardiff: ESP.

Hutchinson, A. and I. Chaston (1994) Environmental management in Devon and Cornwall's small and medium sized enterprise sector, *Business Strategy and the Environment* 3, 1: 15-22

Martin, S. and H. Oztel (1996) The Business of Partnership: Collaborative-competitive partnerships in the development of Business Links, *Local Economy* 11, 2: 131-142.

Moran, Paul (1993) Development of small business support services, in *New Directions in Small Business Research* Roderick Atkin, Elizabeth Chell, Colin Mason eds. Aldershot: Avebury.

Segal Quince Wicksteed (1988) *Encouraging small business start-up and growth: creating a supportive local environment* London: Her Majesty's Stationery Office.

Chapter 11

PARTNERS IN POLLUTION PREVENTION

Increasing Environmental Capabilities of SMEs through Collaboration[1]

THEO DE BRUIJN

PETER S. HOFMAN

1. INTRODUCTION

The concept of sustainability increasingly impacts the nature of industry's operations. Whereas in the past sound economic performance was expected to guarantee success by companies and its shareholders, now increasingly business is led by the so-called triple bottom line. Achieving good economic results needs to be accompanied by minimisation of environmental impacts and increasing attention to social aspects. Over the years, environmental management has therefore developed into a strategic issue for industry. Adopting the triple bottom line implies that environmental management in organisations has to go beyond regulatory compliance and needs to include concepts such as pollution prevention, product stewardship and social responsibility.

Where larger firms have been developing the capabilities needed to achieve the triple bottom line over the last decades, small and medium-sized enterprises (SMEs) often lack the skills, knowledge and expertise, funds and time to make the desired changes. And when they change, it is often

[1]This chapter contains an extended and revised version of an article published in Greener Management International, 30, 2000: 71-82.

T.J.N.M. de Bruijn and A. Tukker (eds.), Partnership and Leadership, 195–215.

narrowed to specific features of the production process or the product by which the firm has earned its place on the market. Therefore, SMEs often have a limited view on the direction of future innovations and tend to tackle environmental issues in an ad hoc manner. Developing partnerships with external parties is seen as a way of overcoming these problems. They can put environmental issues on the company's agenda and bring in technological knowledge, managerial expertise, external and 'unprejudiced' views on the production process, funding via cost-sharing constructions and contacts with investors and banks.

In this chapter we review a specific attempt in developing partnerships, namely the promotion of pollution prevention in the Netherlands. Dutch pollution prevention activities often take place in a collaborative setting. Regional authorities (provinces) play a central role in this by means of their so-called prevention teams. They develop projects in which consultancy agencies, research institutes, municipalities and trade associations collaborate in order to get individual firms active with pollution prevention.

The main question we raise in this chapter is: Are the partnerships effective in delivering environmental capabilities to SMEs? We approach this question from two angles. First, we look at whether companies have been able to improve their environmental performance in the short term. Second, we ask ourselves to what extent the companies have learned how to integrate concepts of pollution prevention into their daily management practices. This second question is particularly important from the perspective of sustainable development. The basic feature of sustainable development is that it concerns an *on-going process* of improvements. Next to a one-off improvement, the continuity of efforts is crucial then. In this chapter we will evaluate to what extent pollution prevention projects have been able to provoke continuous and profound changes within companies.

Our conclusions are twofold: If we look at the short-term effects, we are rather positive about the benefits of the projects and the collaboration they induced. Most pollution prevention projects succeeded in bringing relevant partners together and in realising environmental improvements for the participating companies, although our analysis has also shown that pollution prevention does not necessarily lead to fundamental change in production processes. If we look at the longer-term effects we tend to be more negative. For the majority of companies the pollution prevention project is a one-time experience directed to the generation and implementation of cost-effective measures. Especially more recent projects, with for the sake of efficiency an emphasis on quick-scans rather than on a full-scale analysis with a corresponding improvement program, don't seem to yield lasting effects.

This is partly due to the nature of the set-up of the partnerships in projects and the division of tasks between partners.

Overall, we conclude that most pollution prevention projects do result in better environmental performance by the participating companies, but for the majority it is a one-off experience. There is limited learning regarding the art of pollution prevention itself. Another problem is that current partnerships are not able to voluntarily involve laggards in the pollution prevention projects. To overcome these shortcomings we make some recommendations on how to improve the partnerships and the roles the various partners take on.

The set-up of the chapter is as follows: In the following section we explore recent theoretical and empirical insights into corporate environmental change and the role of pollution prevention. Section 3 focuses on different modes of pollution prevention projects and their relevance to SMEs. Section 4 discusses the results of these projects. Section 5 examines the roles of the various partners and their contribution to processes of environmental change in SMEs. Section 6 gives a conclusion.

2. CAPABILITIES FOR INDUSTRIAL TRANSFORMATION

Initially, policies to urge business towards environmentally friendly behaviour have mostly relied on direct regulation. Although direct regulation has been a powerful tool for adjusting industrial behaviour, its disadvantages soon became apparent: pollution shifted from one media to another rather than being eliminated, it could constrain innovation, enforcement was difficult, it could lead to inadequate or incorrect priority setting and higher transaction costs. Moreover, direct regulation can hardly be expected to answer the challenges of sustainability. Sustainable development is in its core a continuous process of risk assessment in which multiple actors are involved. The issue at stake is therefore not 'compliance to regulation' but how to use the creativity of all the actors involved. Where in the past, specific standards and end-of-pipe technologies prescribed the way companies had to deal with the environment, recent approaches including pollution prevention projects leave much more discretional space to the companies. It is argued that these approaches are more cost effective as new technologies do not have to confront short-term standards that are likely to be sub-optimal in the long term (Ashford, 1996; Harrison, 1999). Central to these approaches is a more conscious and efficient use of resources and a reduction of emissions since it is of crucial importance that environmental problems are being addressed at the source through prevention measures in order to prevent pollution from just being transferred into another form.

According to Hart (1995) pollution prevention is an important step in a development path of companies towards a more sustainability-based strategy. An important premise is that pollution prevention, and the knowledge and capacities it entails, forms a first step in a learning process of companies. A company needs to appropriate certain capabilities in order to develop a successful and competitive sustainability strategy. This specific knowledge and capacity can be acquired through pollution prevention. Lober (1998) views pollution prevention as a process of renewal: "Pollution prevention forces the rethinking of the whole corporate approach to material, energy, and water use and efficiency as well as to pollution control. By focusing on what enters and moves through the production processes, managers can realign their thinking to make production processes use less material and thereby have greater yields while reducing their impact on the natural environment" (Lober, 1998:29). Dieleman (1999) points at the confrontational nature of pollution prevention. He asserts that in order to realise pollution prevention companies need to break through the established routines, conventions, and standard solutions and 'defreeze' established images (Dieleman, 1999:225). Companies need to be confronted with the fact that they overlook the potential for prevention and have developed blind spots for the opportunities that exist.

Hart (1995) argues that in order to realise sustained competitive advantage companies need to develop their environmental management based on the concepts of pollution prevention, product stewardship, and sustainable development. The integration of these concepts into management demands increasingly complex collaborative strategies. Pollution prevention is seen as a first necessary step towards sustained competitive advantage and requires tacit-skill development for the optimisation of resource use and the minimisation of emissions. After acquiring the capabilities for pollution prevention companies will be able to move on to product stewardship. The focus there is on the minimisation of life-cycle environmental costs of products and it requires socially complex skills and the involvement of (and collaboration with) external stakeholders. Sustainable development requires both the minimisation of the environmental burden of a firm's activities from cradle to grave and incorporating social responsibility along product chains. It involves dialogue, consultation, and collaboration with stakeholders regarding company performance and company values at both local and global levels (Hart, 1995). The idea that pollution prevention is a limited but necessary first step in the development process towards sustainable enterprises is widely supported. Roome (2001) for instance states that in order to go from 'compliance' to 'beyond-compliance' companies need to include, among other issues, the notion of pollution prevention. The transition to 'sustainable enterprises' then is a matter of "reinventing the company, together with its relationships with others in the socio-technical systems in which its products/services are embedded". Engaging companies in pollution prevention is therefore of great importance, a first step in a long process.

Over the past decade pollution prevention has evolved into a proven, practical method (Oldenburg and Geiser, 1997:104). The majority of larger companies have now adopted concepts of pollution prevention and environmental management as corner stones of a strategy towards increased efficiency (Hart, 1995; Berry and Rondinelli, 1998). In the earlier eighties, frontrunner companies as Dow Chemical and 3M reported that with the integration of pollution prevention in company policy wastes and emissions have been reduced with 25 to 50 per cent within a period of ten years (Freeman, 1990; Nota, 1991; Porter and Van der Linde, 1995). There is less clarity whether SMEs are able and willing to integrate environmental issues into their management practices (Hoevenagels and Nouws, 1997; Hartman et al., 1999). Since the late eighties pollution prevention has nevertheless become a concept which is extensively used to influence the behaviour of SMEs.

While Hart's framework is predominantly built on the experience of larger companies, Medina-Munoz and Garcia-Falcon (1998) assert that SMEs can gain competitive advantage by developing the concepts of pollution prevention, product stewardship and sustainable development through partnerships at the local level, for example through establishing linkages with communities that aim at conserving the local cultural and environmental heritage. However, environmental initiatives in SMEs, such as the start up of pollution prevention, are frequently impeded by a lack of for instance resources, capacity, and knowledge. It is suggested that partnerships can play an important role in overcoming some of these obstacles. Therefore, over the last few years major attempts have been made to transform SMEs into more sustainable companies by applying a network approach. In essence these attempts aim at improving the organisational and technical capabilities of companies. SMEs may join partnerships with various organisations, such as trade associations, public and private institutions, as well as local business networks. These partnerships' objectives range from implementing environmental management systems, incorporating the concept of pollution prevention, to providing information on environmental business opportunities and technological options (Biondi et al., 1998; Hines, 1998 and in this volume). For example, according to Biondi et al. (1998), small businesses working together in groups find collaboration a useful and efficient way of adopting environmental management systems. This is because SMEs can learn from one another, exploit synergies to reduce their collective environmental impact and develop relations with authorities and other stakeholders in the community.

What can then the networks contribute to the capabilities of SMEs? First of all, they can bring in some of the knowledge that is needed to analyse and evaluate the production process from an environmental perspective. Needed here are technological knowledge, organisational insights and also some legal expertise in order to evaluate what the major environmental issues are and to what extent certain in-efficiencies constitute problems. Second, managerial expertise is needed in order to run a project like pollution prevention. Even if a certain company possesses the necessary knowledge, it may have a hard time to implement a pollution prevention project. External parties may provide an orderly step-by-step approach, ensure participation of key-players within the organisation, provide a reasonable planning, and keep pollution prevention on the company's agenda. Where finances are a problem, projects might also be able to solve some of these issues. The first important financial issue is the project itself. Via cost-sharing companies can often participate against very low rates. The partnerships involved in the pollution prevention projects will have to help and deliver these capabilities to SMEs. Together the collaborating partners must also be able to reach for SMEs in order to get them involved in pollution prevention projects. The general expectation here is that a joint effort will be more effective than isolated attempts.

In sum, the process towards more sustainable production takes place in three phases according to Hart's model. Pollution prevention is the first phase. In many cases SMEs lack the resources to develop new capabilities such as required by pollution prevention. Networks are expected to help out via partnerships. At least in theory, pollution prevention projects therefore seem like an attractive way of pushing SMEs on the path towards more sustainable production. In the next section we will describe how the partnerships actually worked out in our case.

3. PARTNERSHIP NETWORKS IN DUTCH POLLUTION PREVENTION PROJECTS

The Netherlands can be considered a front-runner in the field of pollution prevention. As part of its national environmental policy, prevention activities within companies have been stimulated throughout the last decade. This policy has resulted in projects that differ in the methods used, implementation mode and financial structure. Pollution prevention is usually seen as attractive because it potentially both increases the economic efficiency and reduces the environmental impacts. 'Pollution Prevention Pays' is the slogan that was used to draw the attention of companies to these two aspects.

In pollution prevention projects various organisations are seen to collaborate, such as municipalities, trade associations, universities and consultancy agencies. Below we describe the partnerships that are used to stimulate the introduction of pollution prevention in SMEs in some more detail. *Provinces* play the main role. Almost all provinces in the Netherlands have established a special unit in their organisational structure called 'prevention team'[2]. These teams initiate and co-ordinate pollution prevention projects. Often they also provide the financial resources for projects. Within the context of the *Inter-provincial Consultation* (Dutch acronym IPO) the provinces have jointly developed a common strategy towards pollution prevention, together with *The Ministry of Environmental Affairs* (Dutch acronym VROM). VROM also sets outs long term targets regarding the reduction of waste and emissions and the role of prevention in companies, and supports projects by providing some financial resources (mostly via the provinces). *Intermediary organisations* (e.g. trade associations) provide specific knowledge (such as knowledge on environmental problems and possible solutions for a certain sector of industry), assist in contacting companies and play a role in informing companies and others about the results of projects. Consultancy agencies or specialised research institutes (such as research teams based at a university) often do the actual implementation of projects. They assist companies and carry out some of the tasks needed, for example an analysis of material flows. Other partners include municipalities that can also assist in attracting companies.

[2]Recently, the scope of some of these teams has been broadened. Next to pollution prevention they focus now on introducing elements of industrial ecology and life-cycle management under the heading of 'sustainable production'.

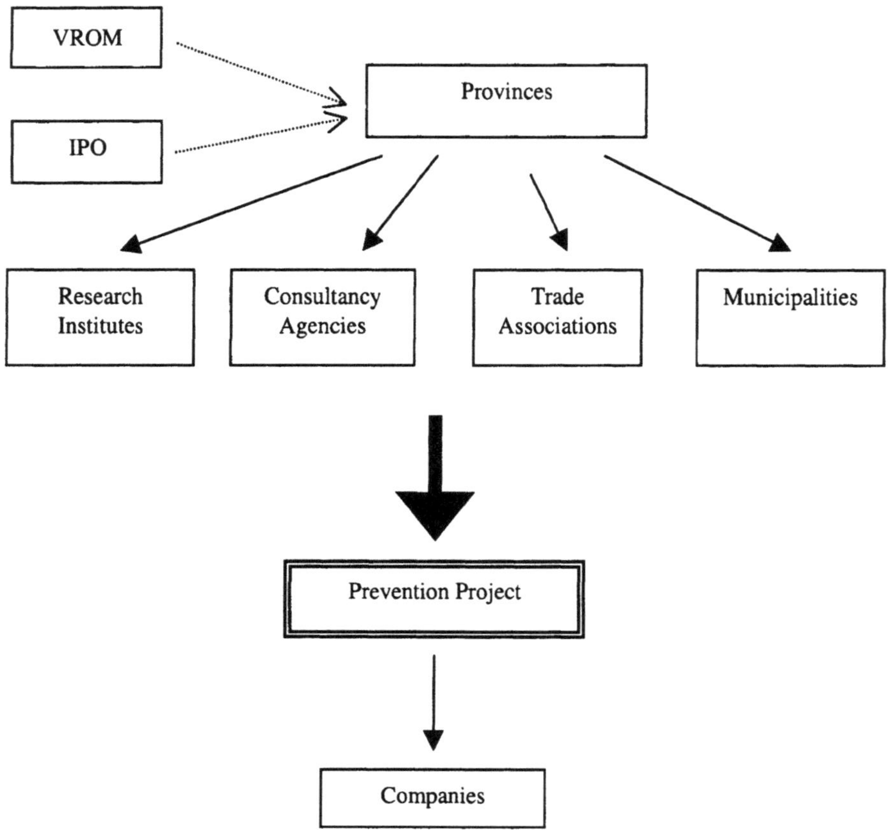

Figure 1. Partnerships in pollution prevention projects

Projects make use of different pollution prevention methods. We may distinguish between two main methods:
- Pollution prevention projects according to the PRISMA-method[3];
- Pollution prevention projects relying on a quickscan.

Below we describe both methods.

Pollution prevention projects according to the PRISMA-method

The PRISMA method was developed on the basis of a Prevention Manual of the American Environmental Protection Agency (US EPA, 1989). It consists of four steps, (1) planning and organisation, (2) assessment, (3) feasibility study, and (4) implementation. It essentially implies a comprehensive analysis of material flows that leads to the generation and implementation of a number of prevention options (Dieleman and De Hoo, 1993). Most of the

[3]PRISMA is a Dutch acronym for Project Industriële Successen Met Afvalpreventie, in English: Project Industrial Successes with Pollution Prevention.

initial pollution prevention projects in the Netherlands made use of this method, sometimes in a slightly moderated form. These projects required an active involvement of the participating companies as they play a significant role in the gathering of data, the analysis of material flows, and the generation of prevention options. Thus, the PRISMA method demands a significant investment in terms of time and money. For example, in the first original PRISMA project the first two stages of planning and organisation and assessment typically took around 30 to 70 days from the company, and around 20 advisory days from the external partner (Dieleman and De Hoo, 1993: 251). In subsequent projects based on these experiences time was managed more efficiently with an average of 20 days spent by companies on the whole project (De Bruijn et al., 1995: 78).

A typical example of a PRISMA project can be found in the wood construction industry. Here a project was initiated in 1992 by a consultancy agency who contacted the trade association and several provincial prevention teams. The goal of the resulting project was to get an insight into possible prevention options within the wood construction sector by using the PRISMA method. During 1993 eleven companies participated in the project. A Project Committee (in which the trade association, the provinces and the consultancy agency were represented) guided the implementation. The actual implementation was supported primarily by the consultancy agency. It participated in the project teams within companies. They saw their role mainly as guiding the process. The company itself had to bring together most of the information and it also had to play an active role in identifying possible options. This cost them around twenty days on average. The number of options differed from 10 to 40 per company; the number of actual measures ranged from 6 to 35. Besides the direct results within the eleven companies the project resulted in a manual for pollution prevention that can be used by other companies in the sector.

At the start of pollution prevention government agencies and companies (mainly the larger ones) could afford making these investments. Later on the call for more efficiency and the involvement of SMEs, with less financial capacity, increased the need for cheaper and less time consuming pollution prevention methods. New methods are based on a quicker assessment of the opportunities for pollution prevention in companies and are less comprehensive than the original PRISMA-method.

Pollution prevention projects relying on a quickscan

The basic idea of the quickscan method is reflected in its name: quickly scanning material flows of a company in order to assess the potential for pollution prevention. Companies participating in these projects have a relatively passive role. In most cases data are gathered through a number of visits by external prevention experts, combined with data provided by the company. This leads to a picture of the overall material flows and an idea of the most promising areas for a focused assessment of possible prevention options. It can also lead to a number of specific options that were relatively easy to identify during the quickscan. It is then expected that companies will further explore additional prevention options. The number of days invested in these projects is considerably lower than in case of a PRISMA project and vary between 0,5 and 3 days (De Bruijn and Hofman, 1998: 22). In most cases the scan is focused on the material flows within companies, but is often accompanied by an examination of the installed equipment so as to assess the technical capabilities of the company. The latter is mostly done in sectors of industry where production techniques are quite similar. Improvements are then partly customised by prescribing state of the art techniques and/or equipment.

An example of a quickscan project can be found in the furniture manufacturing branch. The project was carried out by two consultancy agencies specialised in pollution prevention projects and supported by several provinces. Eighty-eight companies throughout the Netherlands participated in the project after being contacted through mailings and telephone. The prevention quickscan was developed on the basis of PRISMA projects in the furniture branch in two provinces of the Netherlands. The project was set up as follows. Participating companies were asked to provide data regarding their production processes and pollution. After that an advisor visited the company, looked at the production process, added missing data, and went through a list with prevention techniques with the company representative. On the basis of this visit and the provided information the advisor produced a report which was sent to the company. The report was evaluated through either a telephonic interview or a short visit. A supervising committee with representatives from provinces, the furniture manufacturing trade association, and the consultancy agencies monitored progress of the project. The total costs of the project were 120.000 Euros which was funded by the various provinces. Companies participated free of charge in the projects and only had to invest time. It took the companies an average of one day, while the advisors invested around one and a half day per company scan and around thirty days for the acquisition, project management and reporting. On average 8 to 10 options were generated for each company while companies implemented an average 2 to 3 measures. In general these were the most easy to implement measures. Pollution prevention did not have a follow up in most of the participating companies.

On the basis of the evaluation of various pollution prevention projects it becomes clear that project methods have shifted from the (extensive) PRISMA method to less time-consuming quickscan methods. For the external partners this implies other tasks. Instead of only playing the expert role within company-internal project teams and taking care of project management, external partners increasingly do most of the work needed (e.g. the 'analysis' of the environmental performance of the companies involved in the project). After evaluating the results of pollution prevention projects in general, we will discuss the consequences of this shift.

4. RESULTS OF THE PROJECTS

Over the years we have evaluated a variety of pollution prevention projects in the Netherlands. For this chapter we mainly use data from two different evaluation studies of pollution prevention projects. In the first place we refer to an evaluation study commissioned by the Dutch Ministry of Environment. In this project we studied 24 projects carried out in different parts in the Netherlands with the participation of more than 1500 companies (De Bruijn et al., 1995, 1996). Secondly, we make use of an evaluation study of 11 projects carried out under the aegis of the province of North-Holland with the participation of 182 companies (De Bruijn and Hofman, 1998). To evaluate the effects of the 35 projects in total we have interviewed 65 companies (semi-structured, mostly face-to-face, some telephonic). We also analysed all the written material that was available per company. In addition, we talked to consultancy agencies and other research groups who implemented projects, and to provinces in their role as initiator of projects. Besides these empirical data, we also build on several pieces of advice that we have written concerning the future promotion of pollution prevention by provinces (De Bruijn et al., 2000; Arentsen, De Bruijn and Woltjer, 2001).

To evaluate the effects of the projects in companies, we divide the results in (1) direct effects through the identification and implementation of prevention options and (2) long-term effects through the follow-up of pollution prevention projects by companies.

Direct effects of pollution prevention projects

Direct effects refer to prevention options and actual measures. These results can be compared with the costs of the projects (in a broad sense). Among other things, we looked at the total costs of a project and the level of the companies' involvement. Moreover, we looked at the number of participating companies. Table 1 summarises some of the outcomes.

Table 1. Characteristics and Results of Pollution Prevention Projects

Criterion (Costs in Euros[4])	Average score
Number of companies per project	29
Prevention options per company	19
Actual measures per company (implemented options)	6
Costs per project	179,000
Costs per company	6,172
Costs per option	325
Costs per measure	1,028
Companies' involvement in number of days	13

Without some standard for evaluation it is hard to judge these results. Ideally, one would like to compare the costs to environmental reduction data. These proved, however, to be very hard to get. Surprisingly, most of the participating companies could not give exact figures concerning the environmental impact of the project, apart from some fairly rough estimates. When more exact data were available, comparing the various emissions was problematic. Giving a complete overview of environmental effects (i.e. in terms of reduction of waste flows or emissions) nor developing an index for efficiency is therefore not feasible. The supervising committee of one of our evaluation studies judged the costs of 1,000 Euro per measure as "acceptable". At the same time it is obvious that on such a cost basis not all companies can be treated this way. Reducing the costs of pollution prevention projects was considered a high priority.

As noticed before, during the early nineties most projects used the more elaborate PRISMA method. Follow up projects had to become more efficient. Quickscans served that need. This means that the scores for specific projects can deviate considerably from the given averages. For example: PRISMA projects generated 23 options on average, whereas quickscans only generated 15 options on average. And the average cost per measure was over 1,100 Euro in PRISMA projects, while in quickscans this dropped to just under 700 Euro.

The participating companies have generally implemented only a small number of the generated options. The most decisive factor in this is costs. Options are implemented only when they have a relatively short payback period. As said before, it is striking that after options have been implemented, most of the companies cannot give much details on the effects of those measures, both financially and environmentally. However, the general picture is that investments have been paid back through cost

[4]Only the *external* project costs were taken into account. The time companies had to invest is therefore not included.

reductions over a period of two to five years. And the measures have led to lower environmental impacts through reductions in energy use, material use and waste as well as in wastewater.

An analysis of all projects shows that the amount of time invested by the companies in the projects is positively correlated with the quality of the options (De Bruijn et al., 1995:94; De Bruijn and Hofman, 1998:42). Environmental success does not come 'at a bargain'. Good results require commitment, time and money. The projects using the PRISMA methodology therefore, have produced measures of superior quality. They sometimes implied fundamental changes of the production process and the way it is organised. In projects dominated by the quickscan method most implemented options can be characterised as good housekeeping measures.

Below two examples are given that describe the experiences and outcomes of companies who participated in respectively a PRISMA project and a quickscan. *Company A* is an autonomous company, established already in 1935. It produces different styles of furniture for the mass market. The company has some 45 employees. One of them acts as the quality and environment co-ordinator. Company A participated in a PRISMA project in 1993. It invested 25 days themselves in the project. This resulted in some 35 options, of which eventually almost 25 measures were taken. Some examples are:

- ordering of exact sizes of raw material (mostly wooden balks and plates) leading to less waste;
- improved planning of work, resulting in a decreased loss of material. The planning has been fully computerised;
- the instalment of new welding technologies, resulting in less waste material;
- the use of return packages, leading to a decreased use of raw materials;
- the installing of a new paint process, with an optimisation of the paint cabin including the instalment of a distillation process.

The project resulted in positive financial and environmental effects although our respondent could not give exact figures. After the project was concluded, company A has continued to work on pollution prevention. It periodically investigates its material flows and has developed some options.

Company B is a furniture manufacturer, with some 40 employees. In 1996 the company participated in a quickscan project in which it invested only one day itself. The quickscan resulted in five related options that were all realised. It concerns the separation of waste from one to five waste flows. Three flows are now being reused externally (two wood flows, one paper flow). The company is satisfied with the project. Its knowledge of waste flows has been improved and thanks to the project the company could clear some problems with the legal authorities. After the project company B has not generated any new options. For them it was a positive one-off experience. The concept of pollution prevention plays no further role in the company.

Long-term effects of pollution prevention projects

The central idea in pollution prevention projects is to teach companies how to improve their environmental performance by reducing inefficiencies in their material flows. Besides the immediate improvements through the implementation of (some of) the options that come out of the project, it is therefore relevant to look at what companies have learned about pollution prevention. We have investigated whether the pollution prevention projects have had a sustained impact on the behaviour of companies through four main questions:

- To what extent has the project led to increased knowledge of the firm regarding their material flows?
- Has there been a follow up of the project in any kind?
- To what degree does the company make use of the concept or method of pollution prevention that was employed in the project?
- Did the company generate and implement options independently after the project?

Most companies indicated that a positive relationship exists between their knowledge of material flows and the pollution prevention project. For a majority of all companies in the projects there has also been a follow-up after the project. This follow-up takes different forms: from independently and periodically generating options to further research about material flows and waste streams. Although some companies have generated options independently after the project most companies indicated that no further use was made of the method used in the pollution prevention project. The pollution prevention projects seem to have drawn the attention of some firms at least to specific areas where either serious environmental problems exist or can be expected, or to areas that seem very promising (from an economic point of view). The projects however seem to fail in teaching the companies how to deal systematically with the problems and challenges in these areas. Not much is learned about the art of pollution prevention itself.

To investigate this further, we have analysed the level and form of environmental management in participating firms both before and after the project, and have estimated the impact of the pollution prevention project on this. At the beginning of the project most companies had a positive attitude towards the environment and were in the initial stage of implementing some form of environmental management, see Table 2. A significant number of companies were active in addressing environmental affairs and had a partial or fully functioning environmental management system. None of the companies participating in the projects were regarded as defensive with regard to environmental issues. It can be concluded that in general pollution prevention projects do not reach companies that are laggards in environmental management.

Table 2. Environmental management in companies before and after the project

Level of environmental management	Percentage of companies Before projects	in 1998
Defensive	-	-
Beginner	60	40
Active	35	40
Pro-active	5	20

We have also assessed the level of environmental management after the project, see Table 2. About one third of the companies made progress in environmental management as estimated after the pollution prevention project. In a number of cases, the prevention project had played a major part in this (according to our respondents).

Finally, we have looked at investment patterns and the use of information. Most of the companies (90%) indicate that environmental issues have played a role in the shaping of investment patterns. However, none of the companies indicated that there has been a change of course as a result of the pollution prevention project. With regard to information on environmental issues, more than half of the companies indicated that the trade associations were the most important source, while suppliers were mentioned by one third of the companies. The use of information sources has not shifted in companies as a result of the pollution prevention projects.

5. THE ROLE OF PARTNERSHIPS IN POLLUTION PREVENTION

If we evaluate the results presented above, we may conclude that pollution prevention has proven to be a valuable method for improving the environmental performance of companies, because of its primary focus on material flows (where the causes for environmental problems can be found) and its emphasis on minimisation of environmental effects. Most pollution prevention projects did succeed in realising environmental improvements for the participating companies. However, our analysis has also shown that pollution prevention does not necessarily lead to fundamental changes in production processes. For the majority of companies the pollution prevention project is a one-off experience directed to the generation and implementation of options. The project does not result in a changed routine behaviour of participating companies. Thus, the process of renewal and rethinking of corporate behaviour, identified by Lober (1998) as the core of pollution prevention, is often not initiated by the project. Therefore, not much has changed in most companies in terms of environmental management and investment patterns. This is also consistent with findings of Dieleman (1999) who evaluated the effects of pollution prevention in five companies that participated in the original PRISMA projects. He argues that realising more fundamental change is, among other things, dependent on the ability to establish a process of confrontation for all relevant stakeholders within the arena of cleaner production. According to him most PRISMA projects had limited success in changing established routines within the companies (Dieleman, 1999:225-226).

Another problem is that defensive companies are hard to reach through voluntary pollution prevention projects. The participating companies already belong to the relatively active ones. It is fair to question whether these companies really needed the projects in order to make further progress. Maybe they would have pursued in pollution prevention anyway. The most progress can probably be made in laggard companies. The partnerships, however, were not very successful in engaging those companies.

These shortcomings make it doubtful whether the pollution prevention approach will lead to an ongoing learning process towards product stewardship and sustainable development and will be a valuable approach for transforming industry as a whole. To overcome these shortcomings the partnership networks and the tasks need to be re-evaluated. Two basic questions need to be answered:

1. Are the right partners involved?
2. Do these partners carry out the right tasks?

1. The composition of the partnership

The partnerships should exist of organisations that jointly may carry out the tasks identified in section 2. The first task is making sure that the projects are attractive for SMEs. We have asked our respondents who should be the main partner in pollution prevention. The majority of companies named trade associations, with municipalities mentioned in second place (De Bruijn et al., 1995:87). The main partner for SMEs proved to be the trade association because of their longstanding relationship with SMEs and their credibility (De Bruijn and Lulofs, 2000). Since consultancy agencies are the major partners in current pollution prevention projects (during the implementation phase), the partnerships involved need to be reconsidered.

An interesting development concerning companies that are difficult to reach can be found in an approach applied by some provinces in the Netherlands (see Molier et al., 1996). Since a few years certain aspects of pollution prevention, such as an analysis of the material flows, can be included in the requirements for an environmental permit. In the Netherlands most companies fall under the authority of the municipality for their environmental permit. In order to stimulate the municipalities to use this competency, a training program was set up. In this training program local environmental regulators are made familiar with concepts of pollution prevention. The idea is that they will use this knowledge when negotiating with companies about issuing or renewing environmental permits. The first, direct results of this second approach (in terms of measures within companies, costs of the project, etc.) are less positive than the results of pollution prevention projects in which intermediary organisations are the main actors. However, in terms of the kind of companies that are likely to be involved it could prove to be a very interesting approach. Municipalities can take care of the group of laggards since this group nearly always has problems with their permits. More in general, it could be beneficial to include partners in the network that have a relationship with companies based on power (e.g. legal authorities, clients). This more coercive approach can thus facilitate the diffusion of the concept of pollution prevention. Partners that have this kind of power relationships with SMEs can be very valuable if they deploy their influence in the right way. The current partnerships seem to rely too much on voluntary participation, and in some cases had a hard time finding sufficient participating companies.

2. The division of tasks between partners

The second question concerns the content of the projects. Prevention quickscans have become a widespread method for pollution prevention projects. Quick-scans are relatively easy to carry out, are relatively cheap and more efficient than PRISMA projects and often involve more companies per project. However, there is little engagement and commitment among the companies subjected to a quickscan. This also has implications for the number and nature of the options generated and implemented. In projects where companies invest more time in the pollution prevention project the prevention options are more tailored to the specific company and likely to be more profound. The number of implemented options in companies is also higher in these projects. More importantly, a pollution project should be aiming at teaching companies how to integrate pollution prevention in management routines. Current projects, and especially the ones making use of quickscans, are too narrowly focused on realising environmental improvements. External parties often carry out (parts of) the analysis and come up with prevention options without forcing companies to rethink their production processes and its pollution. Although this may seem efficient, we think these projects miss the main goal. If companies do not learn the art of pollution prevention, the projects might just be an inefficient way of getting a one-off improvement.

A comprehensive, PRISMA-like approach that relies on more involvement by the participating companies may therefore seem favourable. In these projects the role of the advisor is not to detect the prevention options but to guide companies in the search towards cleaner production processes. The external advisor needs to focus on integrating the organisational process of pollution prevention in the company and on confronting the company with its routine behaviour. Chances are that companies will sooner enter into a learning process then, but the downside is that it has proved to be difficult to engage especially smaller companies in more comprehensive projects. Pollution prevention projects need to be cost effective without weakening the learning process. Quickscans can be valuable to arouse interest in pollution prevention, but an appropriate follow-up is needed. More in general: Our analysis has shown that often external stimulation needs to be continuous in order to keep pollution prevention on the agenda. Voluntary pollution prevention projects could very well be employed in conjunction with more coercive approaches. Our data have shown that company commitment and involvement in the project is also vital. Otherwise the concept of pollution prevention will not become embedded into the organisation. Sustainable development at least asks for a continuous process of improvement. The contribution of a one-time experience is limited.

According to Hart's model the next step after pollution prevention is product stewardship. In pollution prevention the focus is on waste streams and emissions rather than products and their effects. The product orientation therefore is clearly undervalued in pollution prevention. Also, pollution prevention in current practice is aiming at individual companies instead of provoking co-operation, for instance in the supply chain (Oldenburg and Geiser, 1997:104) or within industrial estates. There are already existing partnerships that can be used. Companies are of course used to collaborate with suppliers and customers. And many estates have their own associations. Future projects should include an analysis of improvements beyond the level of individual firms.

6. CONCLUDING REMARKS

In this article we have taken a close look at the way in which in the Netherlands pollution prevention projects have been conducted over the last decade. Involving a broad range of actors in order to increase the environmental capabilities of SMEs is central to this approach. Our main conclusion is that pollution prevention is at least potentially a powerful method, albeit with some weaknesses. By broadening the scope of pollution prevention and the way it is implemented the results can be improved.

A major problem lies in the continuation of pollution prevention after a project (initiated from outside the company) ends. Our research indicates that in order to move beyond the one time impact of a pollution prevention project, and for companies to use pollution prevention as a concept for environmental management and organisation, it is necessary for companies to enter into a learning process. This implies that companies have to invest significant time in those projects and should not play a passive role. External partners should mainly play an expert role (by providing knowledge and expertise) in this learning process. Furthermore they have a role to play in project management and guiding the search process for prevention. Less focus should be on doing the actual work within companies.

At the same time we know that those kinds of projects are relatively expensive, both for initiators (governments) and the companies involved. An option to overcome this dilemma might be to take advantage of the initiatives, knowledge and capacities of other partners. Often these partners develop their own environmental projects already. We have seen examples of Chambers of Commerce implementing projects with the goal of increasing the awareness of the industry in the region; energy companies implementing projects that aim to increase the efficiency of their customers; municipalities working on the revitalisation of industrial estates, employers associations carrying out a water project, etc. All these projects might result in better effects if co-ordinated. And besides, partners can be of help in reaching companies (especially SMEs), implementing projects, increase the pressure on companies in order to keep pollution prevention on the agenda of those companies. It will also lighten the burden placed on the provinces. Therefore, we feel that besides implementing concrete projects, attention should be paid to the building of a (regional) network or consortium. This consortium can initiate and co-ordinate activities, and thereby generate an ongoing process of environmental change.

What lessons can we draw from this case about the value and prospects of collaboration? We feel that a multi-partnership collaboration can be a potentially effective way of engaging SMEs in pollution prevention. This has great benefits. Not only is pollution prevention a multi-media approach and is it aiming at the reduction of emissions rather than a transfer to another environmental media, it may also be a first step towards a more sustainable production. By bringing together partners with different expertises and skills, some powerful projects can be developed. Our data also show that the partnerships are not always effective though. Two issues seem crucial. First, the set-up of the partnership is important. It needs to include partners that are credible to the target-group and have different relationships with it. Next to partners who may be able to motivate companies, partners who can use some power to force companies, are necessary to reach laggards. Second, projects need to focus both on achieving environmental improvements and on engaging companies into a learning process. Partnerships need to realise that pollution prevention can only be successful in a complicated process of doing, using and interacting. The real value lies in rethinking a companies' production process and its related pollution and not in the implementation of easy options. In the long run it may be more important for companies to learn the art of pollution prevention than just having a one-off experience.

REFERENCES

Arentsen, M.J., T.J.N.M. de Bruijn and J. Woltjer (2001), *Schoner Produceren. Een advies voor een effectieve en efficiënte monitoringssystematiek* (Cleaner Production – an advise for an effective and efficient system of monitoring.) Enschede: CSTM, University of Twente.

Ashford, N.A. (1996) *The influence of information-based initiatives and negotiated environmental agreements on technological change.* Paper for the International conference on The Economics and Law of Voluntary Approaches, Venice.

Berry, M.A, D.A. Rondinelli (1998) Proactive corporate environmental management: A new industrial revolution. *Academy of Management Review*, Vol. 12, No. 2:38-50.

Biondi, V., M. Frei, F. Iraldo (1998) *Environmental Management System Implementation by SMEs: EU Experience and Perspectives.* Paper presented at the Seventh Greening of Industry Network Conference, Rome, Italy.

De Bruijn, T.J.N.M, F.H.J.M. Coenen, K.R.D. Lulofs and N.E. Marquart (1995) *Onderzoek naar de proefprojecten afvalpreventie* (Research into the pilot projects pollution prevention), Ministry of Housing, Spatial Planning and Environment, nr. 1995/20a, The Hague.

De Bruijn, T.J.N.M., F.H.J.M. Coenen and K.R.D. Lulofs (1996) Pollution prevention projects in the Netherlands, *Journal of Cleaner Production*, Vol. 4, No.1:41-53.

De Bruijn, T.J.N.M. and P.S. Hofman (1998) *Evaluatie van afvalpreventieprojecten binnen de provincie Noord-Holland* (Evaluation of pollution prevention projects within the province of North Holland) Enschede: CSTM, University of Twente.

De Bruijn, T.J.N.M. and K.R.D. Lulofs (2000) Driving SMEs towards environmental management; Policy implementation in networks, in *Small and Medium-sized Enterprises and the Environment*, R. Hillary, Sheffield: Greenleaf Publishing.

De Bruijn, T.J.N.M., F.A. Grobbe, P.D. Schultink, P.G. Welman (2000) *Evaluatieonderzoek Preventieprogramma Provincie Noord-Holland* (Evaluation of the Prevention Program of the Province of North-Holland), Hengelo: PMC.

Dieleman, H. and S. de Hoo (1993) Toward a Tailor-made Process of Pollution Prevention and Cleaner Production: Results and Implications of the PRISMA Project, in *Environmental Strategies for Industry*, K. Fischer and J. Schot eds., Washington DC: Island Press.

Dieleman, H. (1999) *De Arena van Schonere Productie* (The Arena of Cleaner Production), Ph.D. Thesis Erasmus University Rotterdam, Delft: Eburon.

Freeman, H. (1990) *Waste Minimisation*, New York.

Harrison, K. (1999) Talking with the Donkey: Cooperative Approaches to Environmental Protection, *Journal of Industrial Ecology*, Vol 2, No. 3: 51-72.

Hart, S.L. (1995) A Natural-Resource-Based View of the Firm, *Academy of Management Review*, Vol. 20, No. 4:986-1014.

Hartman, C.L., P.S. Hofman and E.R. Stafford (1999) Partnerships: A Path to Sustainability, *Business Strategy and the Environment*, Vol. 8, No. 5:255-266.

Hines, F. (1998) *A Model of an Environmental Business Support Network: Evaluating a Partnership Approach in Industrial South Wales.* Paper presented at the Seventh Greening of Industry Network Conference, Rome, Italy.

Hoevenagels, R., J.T.F. Nouws (1997) *Preventie bij bedrijfsmatige activiteiten*, Zoetermeer: EIM/Milieu.

Molier, T., L. Pietersen and F. Verspeek (1996) *Ambtenaren stimuleren preventie? Evaluatie gemeenteproject Noord-Holland*, Amsterdam.

Medina-Munoz, D., J.M. Garcia-Falcon (1998) *Sustainability as a Major Source of Competitive Advantage for Small and Medium Sized Enterprises.* Paper presented at the Seventh Greening of Industry Network Conference, Rome, Italy.

Nota (1991) *Kiezen voor preventie is winnen*, Den Haag.

Oldenburg, K.U. and K. Geiser (1997) Pollution prevention and ... or industrial ecology? *Journal of Cleaner Production*, Vol. 5, No.1-2:103-108.

Porter, M.E. and C. Van der Linde (1995) Green and Competitive: Ending the Stalemate. *Harvard Business Review* 73, 5:120-134.

Roome, N. (2001) *Policies and Conditions for Environmental Innovation and Management in Industry*, paper presented at the International Conference Towards Environmental Innovation Systems, Garmisch-Partenkirchen, September 27-29.

US EPA (1989) *Waste Minimisation Opportunity Assessment Manual.*

Chapter 12

NEITHER SATAN NOR SANTA

Shell, Competitive Advantage and Stakeholders in the Peruvian Amazon

MARILU HASTINGS

1. INTRODUCTION

When Royal Dutch Shell (hereafter 'Shell') re-entered the remote Camisea region of Peru to again plan for the development of a large natural gas field in the area, anticipated effects of the company's activities on local indigenous communities ranged from bringing the evil of destruction and disease to providing prosperity through direct cash compensatory gifts (May et al., 1999a). In fact, in a major break from its past standard operating procedure, the company adopted an unorthodox approach that incorporated extensive stakeholder participation to proactively identify and mitigate the environmental and social impacts of its activities and create net benefits to affected communities *without* providing direct cash payments. Ultimately, Shell was neither Satan nor Santa, but instead, as will be shown, it was a highly competitive company responding pragmatically to external public pressures and concerns.

An exploding number of increasingly sophisticated, well-funded and networked non-governmental organisations (NGOs) around the world are leading the call for global corporations to meet their demands to mitigate negative effects of company activities. In the midst of this socio-political context, firms operating in environmentally and socially sensitive areas in

T.J.N.M. de Bruijn and A. Tukker (eds.), Partnership and Leadership, 217–235.

particular are expected by both host governments and these stakeholders to address these consequences and to play a non-traditional role in providing social services to affected populations (International Finance Corporation, 1998).

In response, some corporations are consulting[1] with their stakeholders to meet these demands, resolve value differences, set priorities, research problems, and implement solutions to identified issues (Long and Arnold, 1995). A variety of theoretical perspectives examine various aspects of collaboration, as reviewed by Wood and Gray (1991). Strategic management theory in particular examines collaborative relationships – from stakeholder partnerships to interfirm strategic alliances – and the impacts they have on a firm's competitive advantage (e.g. Porter and Fuller, 1986; Doz et al., 1989; Khanna et al., 1994). Examples of the positive impacts of stakeholder consultation include reduced investment and shareholder risk, improved innovation as a result of sound employee and contractor relations, increased network of trust-based relationships that in turn provide resources and information, and enhanced reputation and brand value (Svendsen et al., forthcoming).

These benefits suggest that corporate-stakeholder consultation may enhance the sustained competitive advantage of the partner corporation through the development of new intangible resources. This chapter will explore this hypothesis and asserts that, through the process of establishing and managing collaborative relationships between firms and their stakeholders, firms do indeed develop new intangible resources, including new capabilities, enhanced reputation and social capital, which in turn lead to sustained competitive advantage[2]. This argument is elaborated through the analysis of a case study of the stakeholder consultation process developed by Shell prior to commencing natural gas production in a remote area of Peruvian rainforest[3]. The chapter begins with a review of the relationship between stakeholder consultation and competitively valuable intangible resources. The chapter then presents a case study of the stakeholder consultation process created by Shell. The next section analyses the competitive value of Shell's stakeholder integration process. The chapter concludes by outlining management implications and suggestions for further research.

[1]Stakeholder 'consultation,' 'integration,' and 'collaboration' are used interchangeably in this chapter.

[2]While some researchers treat reputation as a component of social capital (e.g. Petrick, 1999), for the ease of discussion reputation and social capital will be discussed separately here.

[3]ARCO and Amoco (now both part of BP) implemented similar programs in Latin America as well. The three cases are analysed together in Hastings (1999).

2. STAKEHOLDER CONSULTATION, REPUTATION AND SOCIAL CAPITAL

Changes in the socio-political environment within which multinational oil and gas firms currently operate, the long history of distrust in Latin American countries of foreign oil and gas companies, and the severe impacts of their activities in sensitive environments has placed the operations of these companies under continuous and increasing scrutiny (Williams, 1997). Companies are often evaluated not only by traditional measures of financial performance, but also by how well they respond to stakeholders' expectations (Prahalad and Hamel, 1990; Waddock and Graves, 1997). Stakeholder consultation involves responding to and integrating stakeholder input into project planning or product design to minimise environmental and social impacts (Hart, 1995). For any oil company operating in a sensitive area, stakeholder integration would likely focus on corporate decisions concerning such complex issues as environmental and resource protection, biodiversity preservation, waste reduction, facility siting, disease prevention, contact minimisation and economic development. Because these issues are complex, solutions are often difficult to determine, and the cost of addressing them can be significant (Hartman and Stafford, 1998), consulting with stakeholders to help ensure sound decisions is essential.

In the process of integrating stakeholders into project design and planning, firms develop new intangible resources – such as new capabilities, enhanced reputation, and social capital – that can create competitive advantage. The resource-based view of the firm argues that competitive advantage is rooted in a firm's resources, which are comprised of its assets, capabilities, information, knowledge, etc. that enable the firm to pursue strategies to improve performance (Wernerfelt, 1984; Dierickx and Cool, 1989; Prahalad and Hamel, 1990; Barney, 1991; Grant, 1991). These resources can be sources of *sustained* competitive advantage if they are valuable, rare, non-substitutable and inimitable (Dierickx and Cool, 1989; Barney, 1991). It is important to note that the new resources developed as a result of stakeholder consultation processes are intangible in that they are difficult if not impossible to assign a specific balance sheet value (Petrick et al., 1999), or trade in the marketplace (Arrow, 1974; Dierickx and Cool, 1989; Hall, 1993)[4]. Petrick et al. (1999) divides intangible resources into either capabilities or assets. Capabilities include the know-how of employees and the collective attributes that comprise organisational culture, while intangible assets include intellectual property rights, contracts and trade secrets and the components of social capital such as trust and reputation.

[4]Tangible resources include plant equipment, land and other physical resources.

Using reasoning similar to that of the resource-based view of the firm, Hall (1993) argues that such intangible resources are integral to the development of competitive advantage.

A number of researchers have recently explored the role of capabilities in establishing competitive advantage and have found that the new capabilities required for stakeholder integration are strategically valuable and lead to competitive advantage (Hart, 1995; Sharma and Vredenburg, 1998; Petrick et al., 1999). Hastings (1999) and Sharma and Vredenburg (1998) examine capabilities, stakeholder integration and competitive advantage in the specific context of oil companies[5]. Using case analyses of three oil companies operating in Latin American, Hastings (1999) found that key strategically-valuable capabilities include environmental management, social responsibility and sustainable development. Sharma and Vredenburg (1998) found the capability for stakeholder integration, higher-order learning and continuous innovation to be valuable in the pursuit of sustained competitive advantage in the Canadian oil industry.

As noted earlier, intangible assets also play a crucial role in developing competitive advantage. The discussion here concerns two competitively valuable intangible assets that result from effective stakeholder integration: reputation and social capital[6].

Social capital is a complex idea with a variety of definitions[7], but can be understood in the broadest sense as an "asset embedded in relationships – of individuals, communities, networks or societies" (Leana and Van Buren III, 1999:538). It is based on social relations between managers and stakeholders, is founded on trust and mutual understanding, encourages co-operative behaviour, and may reduce the need for monitoring (Nahapiet and Ghoshal, 1998; Sharma, 2000). It leads to open communication with stakeholders, it makes possible certain objectives that could not be achieved without social capital or else only at great cost, it facilitates the diffusion of information, it cannot be traded in the marketplace, and it is socially complex (Burt, 1992; Nahapiet and Ghoshal, 1998).

Reputation represents the cumulative judgement of a firm's behaviour and performance over time (Fombrun and Shanley, 1990) and reflects the firm's relative success in fulfilling the expectations of multiple stakeholders

[5]Integrating stakeholders into corporate decision making involves activities and capabilities that are generally beyond the traditional competencies of oil firms, in particular the ability to establish trust-based collaborative relationships with a wide variety of stakeholders, especially those with non-economic interests (Sharma and Vredenburg, 1998). Thus, stakeholder integration requires the development of new capabilities.

[6]Social capital differs from other types of capital in that 'ownership' is shared jointly among those that comprise the social network of relationships and no one party has exclusive ownership rights. Reputation, however, does accrue to only one party.

[7]See Nahapiet and Ghoshal (1998) for a review.

(Freeman, 1984). The related concept of reputational capital refers to that portion of a firm's excess market value that can be attributed to the perception of the firm as a responsible corporate citizen (Fombrun, 1996). Firms send a broad range of economic and non-economic 'signals' that help predict how stakeholders rank the reputation of that firm and one of the most important non-economic predictors is a company's demonstration of social concern (Fombrun and Shanley, 1990). For oil companies in particular, reputation may be especially valuable. Gaining access to new oil fields is a fundamental strategy for continued growth in the oil and gas industry (*Oil & Gas Journal,* 1998)[8]. Access to environmentally and socially sensitive areas is more likely to be granted to companies that host governments trust (Estrada et al., 1997). In this context, reputation with host governments and local communities is perhaps one of a company's most valuable competitive assets.[9] Today, stakeholders are the fundamental source of corporate legitimacy (Freeman, 1984; Hart, 1995; Grolin, 1998) and a company's social performance record, i.e. its reputation, is established through its record over time of effective interactions with stakeholders (Dierickx and Cool, 1989; Waddock and Graves, 1997). Specifically, companies are expected to manage their impacts on biodiversity, indigenous people, colonisation, deforestation, and public opinion (Barker et al., 1997). Given that positive corporate reputation is a competitively valuable intangible resource (Fombrun, 1996), effective stakeholder consultation is thus an important differentiation strategy for oil firms due to its positive influence on corporate reputation and resulting access to oil and gas resources in environmentally and socially sensitive areas.

[8]In April 1998, a study of the ten largest oil companies revealed that one of the group's strategies for continued growth is to focus their activities outside of the U.S. (except for Alaska), primarily in deep water and in areas, such as Latin America, previously off-limits to private capital and technology. Thus, these firms must be in a position to expand operations into sensitive social and ecological areas in developing countries (*Oil & Gas Journal,* 1998).

[9]The global oil industry has a notorious reputation for responding to environmental problems only after a crisis or enforcement of regulations. The industry's dramatic loss in reputational capital is quantified and documented in Bernicke et al. (1997).

3. CASE OVERVIEW[10]

3.1 Introduction

Native Amazonian peoples have long inhabited the remote Camisea region of the Peruvian Amazon forest near the lower Urubamba River, a tributary of the Amazon. The area is characterised by a sensitive rainforest ecosystem and is home to about 10,000 people in 35 villages, representing seven distinct native cultures. These communities vary from isolated groups of subsistence households to more settled, market-oriented villages. Some of these native groups have rarely, if ever, been contacted by people outside of the immediate area. Health conditions in these communities are poor and child mortality is high, despite long-term efforts of religious missions and NGOs to provide health and education services to the region.

Shell first came to this region in the early 1980s during an exploration campaign that led to the discovery of a massive natural gas deposit[11]. During exploration, Shell was accused of abusing human rights, endangering the health of isolated indigenous peoples and disrupting the fragile rainforest environment. In 1996, Shell returned to the lower Urubamba in a joint development venture with Mobil[12]. In response to the legacy left in the 1980s and negative opinion surveys and the Brent Spar and Nigeria crises in 1995, Shell undertook an internal initiative to improve stakeholder relations. Shell management viewed the return to Camisea as a 'laboratory' in which they could test the company's new approach to stakeholder involvement.

Shell adopted a new operating paradigm (Hastings, 1999) comprised of environmental management, social responsibility and sustainable development strategies for achieving its goal of delivering a net benefit to local communities during and beyond the 40-year period in which Shell planned to operate the natural gas facility in the area. Rather than simply making cash payments to the local community to offset negative impacts of operations, Shell favoured community-wide investments, such as health and education programs, that would ensure a wide and long-term distribution of benefits.

[10]This overview is based on the experience of RD/Shell as documented by the Camisea Case Study Team at the University of California-Berkeley's Energy and Resources Group in May et al. (1999a and 1999b). I am indebted to the Team, including Peter H. May, Valeria G. da Vinha, Natan Zaidenweber and Patricia Fernandez-D'Avila for their input.

[11]11 trillion cubic feet of natural gas and 640 million bbls in liquid condensates.

[12]The respective shares of the partners were 57.5 percent (Shell) and 42.5 percent (Mobil) for a project with an estimated cost of $3.5 billion.

To test this new operational paradigm in Camisea, the Shell/Mobil partnership's national operating unit – Shell Prospecting and Development Peru (SPDP) – launched a comprehensive program of stakeholder consultation to incorporate solutions to emerging issues and strategies for investment in social capital and sustainable development into operational design decisions[13].

3.2 Partnerships and Consultation Process

Prior to its return to Camisea in 1996, Shell gathered detailed information about the environmental issues that could affect the Camisea project and prepared an appraisal of the project's likely impacts on the region[14]. Based on the results of this information gathering effort and subsequent stakeholder review, SPDP adopted a consultation strategy and plan for Camisea in October 1995. The plan contained explicit guidelines to actively seek, inform and hear stakeholders affected by the Camisea project. One of the first opportunities for stakeholder input in the project was through the Environmental Impact Assessment (EIA) preparation and associated project design plans. During the EIA process, several key issues and concerns were identified. Among the issues of greatest concern to stakeholders were:

- minimising project effects on uncontacted peoples;
- protecting public health and safety;
- ensuring safe river transport and waste disposal;
- investing in native employment and economic activity;
- avoiding immigration of job seekers into native communities;
- preventing unwanted secondary development; and
- avoiding unsustainable use of natural resources in the region.

A key part of Shell's stakeholder integration effort during the Camisea project was its partnership with Pro-Natura, an international NGO founded in Brazil with an expertise in applying concepts of social capital and sustainable development to corporate-community projects in developing countries. The organisation's principle responsibility in Camisea was to

[13]The project's final design would include three related enterprises: Camisea gas field facilities, gas and condensate pipelines, and coastal distribution facilities in Lima.

[14]Sensitive to the environmental concerns associated with its past operations in the lower Urubamba, Shell fielded a visitation team to the Camisea region in October 1994 to provide detailed information and improve the company's understanding of the environmental issues that could affect the Camisea project. To preempt any problems that could arise, Shell engaged Environmental Resources Management, an environmental consulting firm, in December 1994 to prepare an environmental profile that included an appraisal of the impacts of the proposed gas development project. This report was distributed to the Peruvian government and NGOs as well as international groups.

build internal commitment within SPDP's Health, Safety and Environment department (HSE) for the design of and long-term corporate commitment to regional sustainable development. A condition of Pro-Natura's partnership in the Camisea project was SPDP's continued demonstration of commitment to delivering a net benefit to the local people.

In early 1996, after visiting the project area, Pro-Natura developed a strategy for stakeholder consultations and planning for conservation and sustainable development in the lower Urubamba region. Pro-Natura played a crucial role in the project as both advisor and challenger to the company on a range of social issues. The consultation approach developed by SPDP and Pro-Natura sought to keep all stakeholders informed at all stages of the project. The objective was to work in partnership with a wide spectrum of stakeholders and to include public health and education initiatives as well as steps to minimise or eliminate potential social impacts of operations. Stakeholders were categorised as either 'primary' or 'secondary' to facilitate management of their concerns. Primary stakeholders were individuals and groups directly impacted by the project and secondary parties included those individuals and groups that were indirectly affected by the project, had some influence over matters affecting it, or had a general interest in the project. Consultations involved native community federation leaders and individuals, national and international NGOs, scientific institutions, government agencies and contractors. Consultation with primary stakeholders absorbed the bulk of consultation efforts, particularly those with native communities and their federations.

Two-way communication channels between Shell managers and stakeholders were established through a variety of efforts that included:

- identification and contact with all stakeholders (200 organisations) representing different areas the project would affect;
- provision of thorough and continuous exchange of information to all stakeholders, including thirteen bilingual briefing papers;
- participation in six successive rounds of consultations with over 30 native communities in the lower Urubamba;
- organisation of a series of participatory workshops at the regional, indigenous federation, national and international levels to further facilitate dialogue; and
- incorporation of stakeholder concerns and recommendations to enhance the project design process.

The core of the consultation process was built around the six rounds of meetings with all primary stakeholders. To manage these meetings and to act as the principal interface between SPDP and native communities, SPDP organised a Community Liaison Team of eight Community Liaison Officers.

The liaison team was housed within the HSE department to ensure coherence with the project's integrated socio-environmental strategy[15]. The liaison team's primary responsibilities were to ensure that SPDP's HSE policies and guidelines related to native communities were endorsed and to consult with communities about project developments and provide feedback to SPDP regarding community concerns.

In addition to the rounds of consultative meetings with primary stakeholders, Shell sought to engage stakeholders in project design by helping indigenous federations in the region build the capacity to participate in the consultative process. The federations were legally constituted to represent groups of native communities and to advance their common interests with regard to territorial rights, education, public health, resource management, and economic development. Nearly all communities in the lower Urubamba belong to one of three federations. In 1997, SPDP launched a series of joint meetings with officers of all three federations, other NGOs and government officials. The company also invited federation leaders to national seminars regarding the social and environmental aspects of the project and also paid the travel expenses for officials to travel internationally to visit another comparable gas facility and attend consultation workshops on the project.

Strong participation by NGOs was considered crucial as these groups were viewed as providers of valuable technical and scientific expertise. National NGOs also brought critical information regarding Peruvian social and environmental reality to the process. SPDP invited a number of NGOs to work in partnership with the company in activities related to local communities, biodiversity and monitoring of environmental features of the project itself. The company directly contracted a network of 38 national environmental organisations, the Peruvian Environmental Network[16], to conduct an independent assessment of company activities and to monitor socio-environmental procedures and impacts.

Shell also developed a partnership with the Smithsonian Institution's Monitoring and Assessment of Biodiversity Program to conduct a biodiversity assessment and long-term monitoring plan for the lower Urubamba. The Smithsonian's findings were reported to all stakeholders and used by SPDP, as part of the EIA process, to make informed decisions concerning the project design and operation.

Government agencies were also important partners in the consultation process. Regional governments of the Inka Region and Ucayali Region were responsible for the lower Urubamba and they had unique access to the

[15]Ordinarily, a team of this type would be located within the Operations Department to facilitate communication between stakeholders and project design managers.

[16]*Red Ambiental Peruana* in Spanish.

financing and political legitimacy required to promote long-term regional sustainable development in Camisea. To facilitate the Regions' engagement in long-term actions aimed at sustainable development, SPDP involved the Regions in conducting a diagnostic study of socio-economic conditions in the lower Urubamba. SPDP supported a series of government-co-ordinated workshops with communities that included research expertise from universities and NGOs familiar with Camisea. After completion of the diagnostic study in November 1997, SPDP supported government and local stakeholder meetings to set regional priorities for investments in sustainable development. The company also facilitated visits by government planners to meet with local stakeholders to more clearly define identified local needs.

In conjunction with government authorities, Shell also conducted a health baseline assessment in 1996. The baseline established priorities for developing long-term sustainable improvement in public health, including organisation of health services, basic health education, training of local health promoters, and monitoring disease and pollution. Shell also facilitated boat and helicopter support for government-led vaccination campaigns in local communities and supported meetings between local health workers, government health authorities and local stakeholders. The goal of this approach was to participate in health care improvement without replacing the responsible government agencies.

The final major partner that played a role in the development of the Camisea project design was the international construction contractor hired in 1997 by Shell. The winning consortium included three firms: Bechtel, a U.S. engineering firm; COSAPI, a Peruvian company; and Odebrecht, a Brazilian firm. This consortium (the BCO Alliance) was required by contract to create its own HSE structure to address construction and implementation issues. To ensure that the BCO Alliance followed company policy regarding HSE strategy, Shell placed 22 of its own staff in BCO offices. The Alliance also hired its own CLO team to secure local agreements on the construction arrangements along the pipeline corridor. However, the Alliance approach to consultation eventually involved the more traditional strategy of offering direct compensation to affected communities rather than adopting Shell's innovative approach of developing long-term social capital and sustainable development for the affected communities.[17]

3.3 Results of partnership and consultation process

While Shell's new approach to stakeholder consultation in the Camisea project yielded a rich set of results, the discussion here focuses on three

[17]The particular stakeholder consultation experience of this consortium requires further examination and comparison with the Shell experience.

major issues that could force delays or cancellation of the project if not effectively managed, which Shell identified as 'show stoppers'. These situations provide specific examples of how the company's consultative approach and internal learning process affected project design and development.

No-Roads Approach

One major operational challenge for the project was that of transporting the materials, supplies and equipment required to build the project's massive field facilities in the lower Urubamba. The key issue was field access and whether or not to construct roads. Serious stakeholder opposition to road building in the Amazon demanded an alternative strategy. The company publicly announced a commitment that there would be no roads built in the region early in the project. Operational personnel voiced little objection to the no-roads approach until mid-1997, when internal objections began to be raised against the decision as it applied to in-field roads between the proposed gas plant and seven well clusters. The alternative to roads was the use of transport barges and helicopters to move thousands of tons of equipment and materials. The traditional objections of cost and safety arose from several senior operations managers. However, the HSE team and several other operations managers echoed the concerns of stakeholders that the roads would provide a dangerous attraction to those people seeking timber, land, and a chance to supplement their income with salaried work. Also, breaking the commitment to the no-roads strategy would certainly destroy long-term partnerships already developed with social and environmental NGOs and scientists. At a meeting in January 1998, an intense debate ensued whereupon it was finally agreed that the issue of building in-field roads was a 'show-stopper' and would jeopardise the project. Acceptance of the no-road decision immediately forced design innovations that fundamentally changed many of the underlying assumptions that had previously supported the case for roads. For example, the original well cluster design would have required 1,000 helicopter flights during construction of each of the seven well clusters. These clusters were redesigned to reduce the number of required flights to less than 200. An extensive financial analysis of the no-roads approach was undertaken and showed that this most important and innovative socio-environmental aspect of the Camisea project increased overall costs by less than 0.3 percent.

The Hovercraft

From the start of the drilling campaign, Shell contracted a hovercraft to navigate in shallow river water during the dry season and deliver transport fuel within the lower Urubamba. Operations managers dismissed initial concerns about the effects of the hovercraft. However, another 'show stopper' situation arose when it became clear that the native communities loathed the hovercraft and once shot at it. The issues concerning the hovercraft included the boat's noise, speed, size and wake. Natives were concerned with dangers to their canoes and cargo, as well as to wildlife and aquatic life. In addition, native cosmology led many to believe that the hovercraft was either itself a demon or possessed by a demon. To address these concerns, SPDP sought significant scientific expertise to assess the possible impact on fisheries and game. These assessments found that the impacts would be negligible and these results were communicated to natives in a systematic way. The hovercraft manufacturer designed new propellers that reduced noise levels and remodelled the hovercraft's more intimidating physical features. Shell arranged for red warning boats to precede all hovercraft down the river and hovercraft were required to go slowly past all communities. Hovercraft also underwent an 'inspection for demons' in each community and natives participated enthusiastically in a contest to name each hovercraft. It was also decided that hovercraft would not operate on Sundays, when there was greater river traffic and often increased consumption of alcohol in native communities. A river safety booklet was prepared, approved, and distributed in consultation with local people to cover all hovercraft issues and concerns.

Public Health and the 'Offshore' Strategy

When Shell was first in Camisea in the 1980s, a serious flu outbreak in the native community occurred due to contact with outsiders. The outbreak killed up to half of one native tribe. With this legacy, one of Shell's top priorities was to prevent the introduction of foreign diseases among native populations. Other serious public health impacts had been registered from contact between workers and natives, including prostitution, fathering and abandonment of children, and violence. A recurrence of any of these situations was seen as a 'show stopper'. Therefore, a major operational challenge was to design and implement the Camisea project in such a way to avoid health-related problems. The company devised a two-part solution. First, workers would live at the site and would not be allowed to enter local communities or mix with local inhabitants. No social or economic contact between the workers and the native population was permitted, except for hiring a few dozen local men for field work. Only the members of the liaison team were allowed contact with the local communities and any worker violating this rule was immediately fired. While Shell's commitment to enforcing this rule was initially questioned, the several dismissals that did occur sent a strong signal to other workers and native communities that the company was serious about enforcing this restriction. Second, all workers were required to receive vaccinations prior to entering the region and the company agreed to provide support to local vaccination campaigns. All workers were required to present a 'Health Passport' signed by the corporate medical advisor before an airline boarding pass to the field site would be issued.

4. ANALYSIS

Shell adopted a complex stakeholder integration process to guide its new operational paradigm in Camisea. Stakeholders provided critical input into Shell's decision making about what the most important issues were and how to address them appropriately in the social and environmental milieu of the region. Stakeholder involvement also provided the company with an opportunity to demonstrate its new commitment to environmental and social responsibility in the area, thus beginning the information flow to stakeholders that they can use to form new perceptions of the company and, over time, enhance reputation and build social capital. As discussed earlier in section two, these intangible assets are competitively valuable to the firm.

4.1 Corporate Capabilities

Using Hastings' (1999) framework, several social and environmental elements of Shell's new operational paradigm can be classified into three categories of strategic capabilities – including environmental management, social responsibility, and sustainable development. These three core capabilities are sources of sustained competitive advantage (see Table 1).

First, as described in the Camisea case study, environmental management capability includes the elements that address Shell and stakeholder goals of limiting environmental impacts, including a no-roads approach, 'offshore' strategy, minimising loss of vegetation and biodiversity, and restoration of disrupted areas. Elements of environmental management in the new operational paradigm may reduce costs in the long-term because of reduced risk, the smaller number of personnel maintained in remote locations, the simplicity and lower maintenance required for production equipment, and the lack of road construction and maintenance (Lindstedt-Siva et al., 1996).

Second, social responsibility capabilities refer to those actions that helped alleviate impacts on stakeholders and local communities by involving these groups in project planning, limiting contact with local communities, conducting social impact surveys, and requiring worker vaccinations to prevent the spread of disease to indigenous groups. Social responsibility may pre-empt competitors due to the relationships formed between Shell and local stakeholders. Enhanced community relations and social responsibility by one oil company may eventually raise the expectations of local stakeholders regarding the behaviour of any other oil operator and act as a barrier to entry.

Third, sustainable development capabilities were more complex and involved programs that required long-term thinking and the ability to envision technologies and other innovations that may not exist. The provision of community assistance involved building educational and health facilities (Hastings, 1999). Sustainable development capabilities may help ensure the future competitive position of Shell due to innovative technologies and long-term planning. The development of a new process or approach to oil field operations ahead of its competitors may provide first-mover benefits. When a firm focuses on future industry trends, the company may be in a better position to move quickly to act on business opportunities that arise. The elements of the new operational paradigm and the resulting capabilities and associated competitive advantage are summarised in Table 1.

Table 1. New operational paradigm elements, capabilities and competitive advantage

Strategic Capability	New Operational Paradigm Elements	Competitive Advantage
Environmental Management: minimising pollution and mitigating overall impact on natural areas	• 'No roads' approach • Offshore strategy • Minimise loss of vegetation/biodiversity • Restoration	Lower costs
Social Responsibility: involving local communities from the outset in project planning decisions	• Stakeholder involvement • Limited contact with locals • Impact surveys • Vaccinations	Pre-empt competitors
Sustainable Development: developing strategies and programs to enhance social capital of affected communities while minimising impacts	• Community assistance • Environmentally sound technology • Long-term planning	Future position

Source: adapted from Hastings (1999)

4.2 Enhanced Reputation

Given that positive reputation contributes to the ability of a company to win concessions to oil and gas resources in sensitive areas – where the bulk of new exploration and production activity will take place – reputation enhancement is particularly important. In the 1990s, Shell's reputation had suffered from lacklustre corporate performance, harsh criticism by stakeholders and media exposure concerning the Brent Spar and Nigeria situations, limited demonstration of social responsibility and its legacy of negative impacts during its first foray into the Camisea region in the 1980s. Shell realised that once lost, this reputational capital would be very difficult to reclaim (May et al., 1999b)[18]. As a result, when Shell began operations again in Camisea in 1996, it decided to change the quality of messages regarding its social and moral attitudes, thereby improving the public's perception of the company and, over time, enhancing its reputation. Thus, social pressure and competitive drive for concessions in remote areas motivated Shell to modify its behaviour and adopt a new operational paradigm in Camisea.

In the conceptual framework of Dierickx and Cool (1989), the new operational paradigm elements and capabilities that Shell developed represent flows of sustainability practices while reputation represents an asset stock. Thus, if Shell implements the elements of the new operational paradigm consistently, stakeholder perception will improve and eventually accumulate to provide the company with improved reputation. For oil companies that traditionally suffer from challenges to their reputations and that need to improve their public perception to expand operations into sensitive environments, there is strong incentive to 'build' this asset[19]. For example, a reputation for environmental and social excellence is accumulated (rather than purchased) by consistently implementing a set of sound environmental and social practices over time.

The development of new social capital is another important, and strategically valuable, intangible asset that emerged through the design and process of the stakeholder integration plan. Like reputation, social capital cannot be traded in the market and the nature of the asset can change over time as relationships change. It may increase over time as well, but disappears if the source relationships cease. Unlike other types of capital, social capital is not controlled by a single entity, but is held jointly among all members of a network (Burt, 1992; Leana and Van Buren III, 1999). Thus, in the Camisea case, Shell and the stakeholders jointly control the social capital developed through the stakeholder integration process. Shell, nonetheless, was able to access this jointly held asset and use it to benefit the company. Therefore, Shell earns additional competitive advantage through

[18]According to a Shell study, ten percent of its global profits derived from its reputation (May et al., 1999b).

[19]In older, low-growth industries, reputation is often pre-existing on a number of different dimensions. In the oil industry, for example, damage to one firm's reputation tends to impacts stakeholder opinion of the entire industry (Russo and Fouts, 1997).

the stakeholder integration process by virtue of the development of new social capital resources.

A 'virtuous cycle' appears to be possible as a result of the interaction between Shell and its stakeholders (see Figure 1). Through the stakeholder integration process, new capabilities, enhanced reputation and social capital emerge. As these assets are developed, they feed back into the stakeholder process, which then yields yet higher levels of capability, reputation and social capital, and so on.

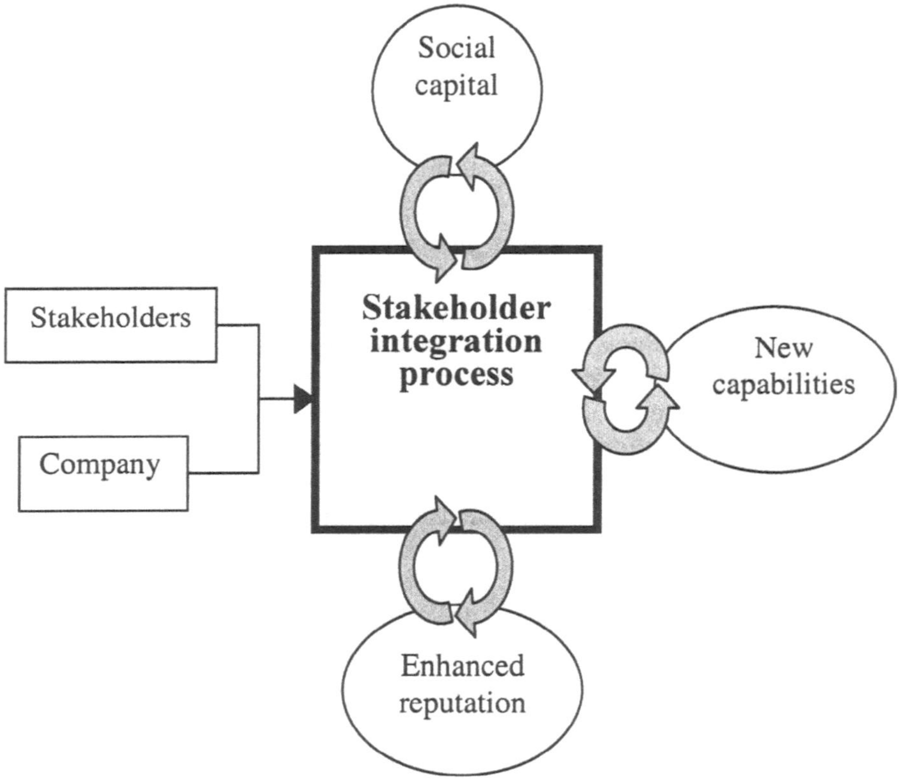

Figure 1. The virtuous cycle

In the long-term, these elements and the reinforcement they lend to Shell's stakeholder relationships may produce ever-increasing capabilities, reputation and social capital. It is conceivable that, as the company's reputation improves over time, it may become a transferable resource that can beneficially influence stakeholder relations in other sensitive areas, thereby improving access to these locations. The new capabilities themselves may reinforce the process directly. As the company develops new ways of improving operations in undeveloped areas, this learning may facilitate the company's ability to continually develop innovative approaches to operations. Through the stakeholder integration process new social capital may emerge and may likewise reinforce the stakeholder process and the relationship itself through the improved levels of trust and empathy between the parties. Social capital, by virtue of its benefits of reducing transaction costs and smoothing information exchange, may reduce the costs and improve the efficiency with which new operational elements are devised.

5. CONCLUSION

The integration of stakeholders guided Shell's project planning in the Camisea region. As shown, the stakeholder consultation process resulted in the development of new capabilities within Shell, the potential for Shell's enhanced reputation, and the development of new social capital that accrued to both parties. All of these intangible resources prove to be competitively valuable for the company.

The strategic importance of stakeholder integration suggests that the pursuit of these collaborations should play a major role in the corporate strategic management process. As stakeholders become more sophisticated and operations move further into sensitive areas – from the tropics to the arctic – companies may find it to their strategic advantage to involve stakeholders proactively in project planning.

Interesting avenues of continued research abound regarding the dynamic relationship between the elements of stakeholders, companies, trust, reputation, social capital and sustainability. An analysis of social capital as a precursor to corporate sustainability would contribute to the understanding of why and how proactive companies pursue sustainability strategies. A review of trust as both an antecedent to and a result of social capital, reputation enhancement and sustainability would be useful. The notion of a 'virtuous cycle' involving intangible assets and capabilities as both inputs and outputs of stakeholder integration processes would be valuable.

ACKNOWLEDGEMENTS

I am indebted to my colleague at the Mitchell Center, Mitchell L. Mathis, for his valuable editorial and substantive comments on this chapter and lively discussions of the virtuous cycle. I am also grateful to Peter May and his colleagues on the Camisea Case Study Team for their excellent documentation of the Shell project in Peru. The analysis on this important corporate experience would not be possible without their efforts.

REFERENCES

Arrow, K. (1974) *The Limits of Organisation.* New York: W.W. Norton & Company.

Barker, G., G.R. Smith, F.J. Vacas, E.K. Swingholm, R.M. Yuill, M.A. Aleman (1997) Managing non-technical risks associated with seismic operations in the tropical rainforests of Ecuador. *Oil & Gas Journal Special Issue*, 50-60.

Barney, J.B. (1991) Firm resources and sustained competitive advantage. *Journal of Management*, 17: 99-120.

Bernicke, J., G. Dowell, S. Hart, A. Sastry (1997) *Corporate Environmental Reputation: Asset or Albatross?* Paper presented at the Corporate Reputation, Image, and Competitiveness conference. January, New York.

Burt, R.S. (1992) *Structural Holes: The Social Structure of Competition.* Cambridge, MA: Harvard Business School Press.

Dierickx, I., K. Cool (1989) Asset stock accumulation and sustainability of competitive advantage. *Management Science*, 35:1504-1513.

Doz, Y., G. Hamel, C.K. Prahalad (1989) Collaborate with your competitors and win. *Harvard Business Review*, 67:133-139.

Estrada, J., K. Tangen, H.O. Bergesen (1997) *Environmental Challenges Confronting the Oil Industry.* New York: John Wiley & Sons.

Fombrun, C., M. Shanley (1990) What's in a name? Reputation building and corporate strategy. *Academy of Management Journal*, 33(2):233-258.

Fombrun, C.J. (1996) *Reputation: Realizing its Value from Corporate Image.* Cambridge, MA: Harvard Business School Press.

Freeman, R.E. (1984) *Strategic Management: A Stakeholder Approach.* Boston: Pitman.

Grant, R.M. (1991) The resource-based theory of competitive advantage: implications for strategy formulation. *California Management Review*, 33:114-135.

Grolin, J. (1998) Corporate legitimacy in risk society: the case of Brent Spar. *Business Strategy and the Environment*, 7: 213-222.

Hall, R. (1993) A framework linking intangible resources and capabilities to sustainable competitive advantage. *Strategic Management Journal*, 14:607-618.

Hart, S.L. (1995) A natural-resource-based view of the firm. *The Academy of Management Review*, 20:986-1014.

Hartman, C., E. Stafford (1998) Crafting enviropreneurial value chain strategies through green alliances. *Business Horizons* March-April, 41:62-72.

Hastings, M. (1999) A new operational paradigm for oil operations in sensitive environments: an analysis of social pressure, corporate capabilities and competitive advantage. *Business Strategy and the Environment*, 8:267-280.

International Finance Corporation (1998) *Doing Better Business Through Effective Public Consultation and Disclosure.* Washington, D.C.: International Finance Corporation.

Khanna, T., R. Gulate, N. Nohria (1994) Alliances as Learning Races, in *Academy of Management Proceedings,* D. Moore ed.; 42-46.

Leana, C.R., H.J. Van Buren III (1999) Organisational social capital and employment practices. *Academy of Management Review*, 24(3): 538-555.

Lindstedt-Siva, J., L.C. Soileau IV, D.W. Chamberlain, M.L. Wouch (1996) Engineering for Development in Environmentally Sensitive Areas: Oil Operations in a Rainforest, in *Engineering Within Ecological Constraints* P.C. Schulze ed. Washington, D.C.: National Academy Press.

Long, F., M. Arnold (1995) *The Power of Environmental Partnerships.* Fort Worth: The Dryden Press.

May, P., A.H. Barbosa, N. Zaidenweber, P. Fernandez-Davila, V. Goncalves da Vinha (1999a) *Corporate Roles and Rewards in Promoting Sustainable Development: Lessons and Guidelines from Camisea,* Research Report, ERG-99-1, Energy and Resources Group, University of California at Berkeley.

May, P., A.H. Barbosa, N. Zaidenweber, P. Fernandez-Davila, V. Goncalves da Vinha (1999b) Royal Dutch Shell Sustainability Strategy, in *Corporate Incentives and Environmental Decision Making* M. Hastings ed. The Woodlands, Texas: Houston Advanced Research Center.

Nahapiet, J., S. Ghoshal (1998) Social capital, intellectual capital, and the organisational advantage. *Academy of Management Review*, 23(2): 242-266.

Common Financial Strategies Found Among Top 10 Oil and Gas Firms. *Oil & Gas Journal* 1998, 96(16): 27-33.

Petrick, J.A., R.F. Scherer, J.D. Brokzinski, J.F. Quinn, M.F. Ainina (1999) Global leadership skills and reputational capital: intangible resources for sustainable competitive advantage. *Academy of Management Executive*, 13(1): 58-69

Porter, M., M.B. Fuller (1986) Coalitions and Global Strategy, in *Competition in Global Industries.* Boston: Harvard Business School Press.

Prahalad, C.K., G. Hamel (1990) The core competence of the corporation. *Harvard Business Review*, 68(3): 79-91.

Russo, M., P. Fouts (1997) A resource-based perspective on corporate environmental performance and profitability. *The Academy of Management Journal,* 40(3): 534-559.

Sharma, S., H. Vredenburg (1998) Proactive corporate environmental strategy and the development of competitively valuable organisational capabilities. *Strategic Management Journal,* 19: 729-753.

Sharma, S. (2000) *Stakeholder integration and corporate sustainability strategy: a dynamic capability perspective.* Houston Advanced Research Center working paper.

Svendsen, A., R. Abbott, R. Boutilier, D. Wheeler (2001) Stakeholder relations: pathways to competitive advantage (draft). *CA Magazine,* August 2001.

Waddock, S.A., S.B. Graves (1997) The corporate social performance-financial performance link. *Strategic Management Journal,* 18(4): 303-319.

Wernerfelt, B. (1984) A resource-based view of the firm. *Strategic Management Journal,* 5(2): 171-180.

Williams, B. (1997) Foreign petroleum companies developing new paradigm for operating in rainforest region. *Oil & Gas Journal Special Issue*: 37-42.

Wood, D.J., B. Gray (1991) Collaborative alliances: moving from practice to theory. *Journal of Applied Behavioral Science,* June 27(1):3-22.

Chapter 13

LESSONS FROM A BROKEN PARTNERSHIP

The Case of ARET

MARIE-FRANCE TURCOTTE

BASMAH ALI

1. INTRODUCTION

This chapter examines the outcomes of a Multistakeholders Collaborative Roundtable (MCR) in the context of environmental problems. The case studied is the multistakeholder initiative ARET (Accelerated Reduction/Elimination of Toxics), a voluntary non-regulatory program, which aims to achieve virtual elimination of emissions of 30 persistent, bio-accumulative and toxic substances (PBTs), and the reduction of another 87 toxic substance emissions to levels insufficient to cause harm. ARET grew out of a proposal from leading industry executives and environmentalists (The New Directions Group) to the Canadian Minister of Environment in late 1991.

Multistakeholder Collaborative Roundtables (MCR) have often been presented as new mechanisms of co-ordination between various stakeholders and economic organisations for problem-solving at the societal level and especially in order to achieve an ecologically sustainable economy. The environmental awareness corresponds to a recognition of the interdependence between societies and the environment, and between ecological and economic systems. Interdependence calls for a co-ordination effort at multiple levels and among multiple stakeholders. The two best-

T.J.N.M. de Bruijn and A. Tukker (eds.), Partnership and Leadership, 237–250.

known co-ordination mechanisms at the societal level, the market system and the regulatory system, have shown evidence of many failures in regard to environmental problems. It is in this context that co-ordination efforts through multistakeholder forums, partnerships, negotiations and collaborative planning were called for by the Brundtland report in 1987, at the Earth Summit in 1992 and by numerous academics (e.g. Barouch, 1989; Kellman, 1992; Callon, 1993; French, 1995; Long and Arnold, 1995; Porter and Salvesen, 1995; Healey, 1997; Roome, 1998a and b; Clarke, 1998; Clarke and Roome, 1999; Hoffman, Gillepsie, Moore and Wade-Benzoni, 1999).

Furthermore, some have perceived MCR as a mechanism of self-regulation for an industry. Indeed, it has been suggested that, instead of tinkering with public institutions, society should look to private sector institutions to undertake environmental protection initiatives, under the rubric of what have become known as "voluntary pollution prevention initiatives" (VanNijnatten, 1998).

The chapter presents the case study of ARET. This case started as a voluntary, multistakeholder initiative including representatives of governments, industries and NGOs. The NGOs however, eventually left. The results of the present and previous case studies show that self-regulation of industry should not be expected as a direct outcome of such a multistakeholder initiative. It can however, contribute to the achievements of environmental objectives by industry members.

2. METHODOLOGY

The research strategy was that of the case study and particularly that of the embedded case study, i.e. including many units of analysis (Yin, 1994). The units of analysis corresponded to the issues of debate between the participants during the studied MCR. The units of analysis corresponded to 'hybrid objects' (Latour, 1991) that evolve through the debate in a 'hybrid forum', that is to say, through stakeholders' negotiations in an MCR. By choosing such units of analysis, the research distinguishes itself from most studies on MCRs that tended to concentrate on the different stages of the process and on the conditions facilitating consensus without describing in detail the MCR outcomes. Units of analysis were chosen with an inductive process after data gathering.

The MCR observed for the case study was ARET, a Roundtable organised at the national level by the Canadian Minister of the Environment in 1991 with the intent to accelerate the reduction and elimination of toxic substances. Data were gathered by in-depth interviews (8 respondents for a total of 10 hours), document analysis as well as available data study (Leiss, 1996; VanNijnatten, 1998; Gallon, 1997). Available data analysis are often the preferred methodology for reasons of time and cost (Kervin, 1992:353), and in our specific case allowed to assess the perspective of participants we were unable to reach. The data were analysed by means of an inductive process (Strauss and Corbin, 1990). In the first analytical step, data were attentively read and annotated with marginal remarks (Miles and Huberman, 1994). With the help of the software 'Hyper Research', the marginal remarks were gathered in themes, and then synthesised and systemised into files. Triangulation of data sources allowed for validity (Yin, 1994). On the base of the systemised and triangulated data, the case was written.

3. THE CASE OF ARET

What is ARET? ARET (Accelerated Reduction/Elimination of Toxics) is a voluntary non-regulatory program, which aims to achieve virtual elimination of emissions of 30 persistent, bio-accumulative and toxic substances (PBTs), and the reduction of another 87 toxic substance emissions to levels insufficient to cause harm. When and how did ARET start? ARET grew out of a proposal from leading industry executives and environmentalists (The New Directions Group) to the federal Minister of Environment of Canada in late 1991. The New Directions Group proposed a co-operative approach to first identify, then reduce or eliminate the most significant toxic substances. The proposal suggested that stakeholders work together to develop a framework for action on toxics that would be faster and more effective than relying on regulations alone.

The minister responded by launching, in 1992, a group known as the ARET Stakeholders Committee. The Committee originally included representatives from industry, health and professional associations, federal and provincial governments, and aboriginal, environmental and labour groups (Environmental Leaders 1, 1994). The ARET stakeholders committee had three main objectives: (1) to establish criteria for defining toxicity; (2) to compile a list of target substances based on these criteria; (3) to devise a means by which industry could address its toxic emissions (Environmental Leaders 1, 1994). By late 1993, the committee finished its work on identifying, assessing and categorising a list of 117 toxic substances for immediate action. This allowed ARET to secure a substantial list of targeted substances that was based on multi-stakeholder consensus (Leiss, 1996:11). For all our respondents, establishing categories and listing of substances was one of the most successful learning of ARET. It made them knowledgeable

Table 1. ARET's historical events

1991	Leading industry executives and environmentalists propose a co-operative approach to identify and reduce or eliminate the most significant toxic substances.
1992	The Federal Minister of Environment of Canada creates the ARET Stakeholders Committee.
1993	The ARET Committee finished its substance identification, assessment and categorisation, and presents a list of 117 toxic substances for immediate action. Environmental NGOs and unions' representatives leave the ARET Committee because there was no consensus on several issues pertaining to the further mandate of ARET.
1994-97	The remaining ARET members resume their activities to reach their objective to reduce the use of the toxic substances from the list.
1998	The ARET Committee reports that many of the year 2000 targets have already been met in terms of substance reduction.
1999	A renewal process for ARET is initiated.
2000	Workshops are organised to discuss the issues and design of a renewed ARET program. Among the issues addresses is the perceived need for NGOs to come back to the ARET process.
2001	The final report of the current ARET program should be released detailing the achievements made from the base year to 2000.

of what substance to go after, as well as its level of toxicity, and bio-accumulativeness.

> "I would emphasise, however, because it is often not appreciated, despite the break-up of that group, the job it did in screening over 2,000 substances and classifying them into one of four lists is a piece of work that we have been able to draw from beneficially ever since. So I would, even though the group broke up, not want to regard it as a failure, because of its' achievements on that front." (Interview with a respondent)

However, the committee could not reach consensus on the further mandate of ARET and on the role the federal government should play in it. Because of these disagreements, in September 1993, NGOs (including Pollution Probe, the Canadian Labour Congress, the Toxics Watch society of Alberta, the West Coast Environmental Law Association, Great Lakes United and Union québécoise pour la conservation de la nature) and Labour organisations participating on the ARET Committee withdrew from the ARET process.

Our respondents agreed that the NGOs were expecting something more of the initiative. In a letter send to Environment Canada's Assistant Deputy Minister Clarke on September 17, 1993, NGOs and Labour groups described the ARET initiative as "a complex and incredibly cumbersome initiative". The disagreements pertained to four main issues: (1) the regulatory versus voluntary nature of the program; (2) elimination versus reduction of the targeted substances; (3) targeting exclusively outdoor emissions versus also including indoor emissions; (4) the role of the government or its perceived lack of leadership. In the following, these issues are discussed more thoroughly.

NGOs and Labour groups had expected the ARET process to lead to "a legislated program directed toward eliminating the most hazardous of substances" (letter to Environment Canada's Assistant Deputy Minister Clarke on September 17, 1993). However, the other participants wanted ARET to lead to a voluntary program directed to reducing the out-of-plant emissions of the most hazardous substances.

Some environmental groups anticipated that the program would be regulatory in nature; it wasn't even clear that it was going to be an entirely voluntary program (Interview with a participant to ARET representing a governmental agency). Remaining participants to ARET perceived the voluntary approach in ARET to be a preferable alternative than regulation for a number of reasons ranging from giving them flexibility to deal with environmental problems, to cutting costs, and saving time.

> "I think if a company or a source is aware of the issue related to its process, they are in a much better position to take responsibility for it and design their own solution than having someone impose upon them in a general way and then it gets much more expensive and much more difficult." (Interview)

Why would companies join a voluntary program such as ARET? Answers to that question were linked by our respondents from the industry sector to the notion of corporate social responsibility, which in turn is perceived as being related to customer satisfaction and good management practices. Institutional pressures, more specifically some sort of isomorphism to the business environment, were also cited to explain participation in the ARET voluntary program.

> "And I think it just makes sense for anybody who wants to build anything to make it in harmony with its environment. But not only the environment like a few plants and trees but as well the communities living around...Environment in general is the way to operate and to live." (Interview)

> "The way we got involved I would say it was more at an industry level. Nobody was pushed to be part of the program; it was all on a voluntary basis. But, since the neighbour is doing it, let's do it as well! So, that's how it was done. We signed. I'm not sure we were leaders in the industry as a company...but it's more or less a general movement from the industry." (Interview)

The remaining ARET participants had joined ARET because they believe in the voluntary approach, preferring it to governmental regulations. Still, the participants expressed their perception that ARET, as a voluntary initiative, needs mechanisms to ensure compliance by its participants. In a consultation paper on the future of ARET submitted by Environment Canada (1999), it calls for vigorous administration and enforcement of regulations to act as powerful motivator for voluntary programs, including ARET. About the importance of compliance mechanisms and regulation to achieve environmental objectives, the following quotes are revealing:

> "We did changes but it's not because of ARET. It's because of something else: regulations! The big one was the new federal and provincial affluent regulation that required us to have secondary treatment in place at each mill. And by doing that, you answered some of the problems that were raised in ARET. But ARET didn't trigger that...But from the industry side of it, it's more the regulation that triggered us to do something than ARET." (Interview)

> "One of the problems is that there is a lack of ways of verifying the emissions and reduction of the emissions that they report are actually true. And it's not just a weakness of voluntary, it's a weakness for regulatory as well... Because unless you have an enforcement officer measuring every day...it could be misreporting and there's not enough enforcement to be able to say accurately every single emission that is reported is correct." (Interview)

The critics of the voluntary approach also raise the problem of the free riders. The free riders notion was new for some participants, as they didn't have to deal with it either before or in another similar initiative. Free riders are those companies that derive benefits from the program without incurring any of its costs. They are a number of companies within different industries that refuse to participate in the initiative, or will sign on to participate and then do too little or nothing to meet their commitment (Gallon, 1997:10). This results in an uneven playing field where some of the participating companies are committing resources to meeting voluntary targets and others refuse to commit similar money and staff time. Free riders also benefit from the credible image of environmental programs, such as ARET, when in reality they are not participants in them.

Two hundred and seventy eight plants joined the program while two hundred and fifty have not. Two sectors seemed not to have any free riders among their members, that is the Aluminum sector and the Chemical Producers, who are part of the Canadian Chemical Producers Association sector. Both have a 100% participation in ARET. Still, even these sectors expressed their frustration from free riders from other sectors, and from free riders of sectors not participating in ARET at all, and blamed it mainly on the voluntary nature of the program and lack of governmental authority in it.

If there is agreement among the participants to recognise that free riders can jeopardise voluntary initiatives, there are mixed views on the extent of the free riders problem in the specific case of ARET. For one of our respondents, free riders to ARET were not a central problem. On the other hand, critics of the voluntary approach present numbers that would let think that free riders are a bigger than marginal problem to ARET. In the Pulp and Paper sector, 36% of the 59 members of its association (CPPA) would refuse to participate in ARET (Gallon, 1997:10). In the Oil and Gas sector, only four of the eight members of the Canadian Petroleum Products Institute (CPPI) participate in ARET. They only represent 60% of Canada's total refining capacity (Ibid.). It is perceived that the issue of free riders could become increasingly important under a new ARET as substances are added to the program and the costs of emissions reductions/elimination measures mount (Environment Canada, 1999).

The second main issue of disagreement among initial participants to ARET pertains to the elimination as oppose to the reduction of use of the targeted substances. The NGOs believed that the original purpose of ARET was to "identify the most hazardous toxic substances and then develop strategies for their phase-out". The NGOs also believed hat only 'hazard-criteria' should be used in determining action on candidate chemicals and not the 'risk assessment' proposed by industry, which would have led to the ranking of toxic substances according to their 'relative' risk rather than their chemical characteristics. For the NGOs, there were "no safe levels" and elimination was the "only appropriate long-term strategy". This meant uncompromising attempts to reduce both the generation and use of persistent toxic chemicals (Leiss, 1996:12). On the contrary, the participants to ARET representing the industry perceived that the appropriate mandate for ARET should be the reduction of the emissions of the targeted substances. The following quote from a respondent represents this point of view.

> "The environmentalists were interested in not using things, and were getting to agree to prevent using substances. While our view was that substances that were important to use could be used but to minimise the releases to no harm levels, so we should continue using them or switch to a better chemical with less harm and less cost because customers would like the companies to do so." (Interview)

The third main issue of disagreement among initial participants to ARET pertains to the extent of ARET's mandate: should it cover exclusively out-of-plant emissions or also include in-plant emissions? Labour and NGOs wanted to address both "in-plant and out-of-plant releases" as both have the potential to harm the environment and humans (Leiss, 1996:12).

> "Labour of course lost all interest when the program was oriented towards releases and not safety in the workplace and use of toxic chemicals in the workplace so they walked out." (Interview)

The fourth main issue of disagreement among initial participants to ARET pertains to the role the government should play in ARET and, more specifically the perceived lack of leadership from Environment Canada. In a letter sent in 1993 to Deputy Minister Clarke, Environment Canada, the NGOs cited that among their reasons for leaving the multi-stakeholder process is "the lack of leadership shown by Environment Canada". On the NGOs' side, this perceived lack of leadership was associated with the fact that ARET turned out not to be a regulatory program. On the industry's side, it was perceived that the government should be more proactive to encourage voluntary participation to ARET.

> "The solution is that the government should be withholding taxes, put a bit of money to encourage companies to join on a voluntary basis". (Interview with a representative from the industry sector)

Participants from both the NGOs and the industry sectors perceive a lack of leadership from the government because it plays 'only' a role of stakeholder. This seems paradoxical because, in principle, all the participants to a multistakeholder consensual process should have equal status and there should not be a hierarchical structure. This suggests that innovation and learning is needed to redefine the notion of leadership within a multistakeholder context and within a voluntary program.

Among the critics of the government's role within ARET, Versteeg (1998) suggested that the possible activities that the government could play as a leader included: setting targets, establishing regulations, ensuring a level playing field, co-ordinating international environmental agreements that influence and could be influenced by ARET, verifying the process and the results, promoting fiscal measures, incentives and rewards, providing government and political endorsement, dealing with free riders, protect the integrity of this 'national' and 'cross-provincial' program. Environment Canada (1999) suggests a range of financial and non-financial incentives to be offered, both positive and negative, by the government to motivate industry action in voluntary initiatives. Positive incentives could include: regulatory burden relief, regulatory and policy stability, technical assistance, public recognition of good performance, and financial incentives in form of R&D support, and investment and tax credits. Other negative incentives could include: a credible regulatory threat, adverse publicity, and taxes.

After the NGOs left the table in 1993, the remaining members chose to continue with the ARET collaborative process. So, since then, the ARET committee contains representatives of governmental departments, and representatives of the different industry sectors, and carries on with the maintenance of the collaboration process among participants. Yet, any changes to be made to the mandate of ARET, needs the grouping of the initial multi-stakeholders' committee to do so.

ARET's long-term goals were the virtual elimination of emission of 30 persistent, bio-accumulative and toxic substances and the reduction of another 87 toxic substances to levels insufficient to cause harm. ARET's short-term goal for the year 2000 were to reduce persistent, bio-accumulative and toxic substance emissions by 90 percent and all other toxic substance emissions by 50 percent. In early 1994, the ARET stakeholders committee challenged selected companies, particularly in the resource and manufacturing sectors, as well as government departments and agencies to voluntarily reduce or eliminate emissions of ARET substances to achieve specific targets by the year 2000. Participants were asked to state their commitments in publicly available action plans. Currently, 316 facilities from 169 companies and government organisations have submitted action plans to the ARET program (Environmental Leaders 3 -Update-, 2000).

Since the short-term goals of ARET were set for the year 2000, a renewal process for ARET has been initiated. In September 1999, the ARET Secretariat released a discussion paper on the future of the ARET program. It presented options, and key issues and principles to be considered when designing a successor program to ARET. The issues presented include increasing participation in ARET, evaluating the current ARET substances list with the objective of adding and deleting certain substances based on scientific data, and verifying the data reported by ARET participants. In late 2001, the final report of the current ARET program will be released, detailing achievements made from the base year to 2000 (ARET's Environmental Leaders 3 Update, 2000).

There is a consensus among the stakeholders interviewed that the NGOs left a sense of vacuum when they left, and that ARET is missing public credibility because of that. They all agree that a new multi-stakeholder process, including NGOs, is needed to discuss the future of ARET2. NGOs' most important role is probably in legitimising the approach in the eyes of the public and giving it credibility (VanNijnatten, 1998).

> "It's unfortunate that in the absence of all environmental organisations, ARET has not been perceived as credible...it will never be perceived to be credible no matter what we do until there are environmental groups participating in ARET." (Interview)

> "I was pleased to see the significant industry desire to have the NGOs back at the table. In fact, I see NGOs involvement as a key element of helping industry

build the business case for ARET2....Thus, before ARET2 is launched, a concerted effort is needed to interest NGOs in participating." (Pollution Probe in a letter to Environment Canada, September 13th, 1999)

4. DISCUSSION AND CONCLUSION

The ARET case is a prominent example of a voluntary initiative that initially involved multistakeholder participation with representatives from the NGOs, industries and governments. It provides many lessons pertaining to consensus-based decision-making within a multistakeholder context, governance, learning and legitimacy.

In the ARET case, the initial participants did achieve consensus on lists of targeted substances but did not reach consensus on what do about it. Should the use of these substances be eliminated or reduced? Should they be dealt with regulations or through a voluntary program? Not only the initial participants could not reach consensus on these issues but it lead to the break-up of the initial multistakeholder group when the NGOs and the labour associations withdrew.

Considering the break-up of that initial partnership, one might ask why did it go wrong? As suggested by a rationalistic perspective, one might expect that when placed within an appropriate discussion forum, people should be able to speak to each other, to understand each other and to find consensus. The most rational arguments should convince every participant. Furthermore, this should provide the opportunity to realise that earlier conflicts were often caused by misunderstandings. In this regard, the following oranges' story is often presented as a case in point. Two groups of people need 10 kilograms each of oranges to produce two different types of medications. There are only 10 kilograms available and each group need the totality of this quantity so that the medication they need for their survival can be produced. Apparently facing a cruel dilemma, representatives of each group met within a rational collaborative process. Participants to this multistakeholder (two groups) forum eventually discovered that one group needed the flesh of the oranges and the other group needed the peelings.

In the ARET case, the initial participants did find some common grounds, for example the established lists of substances to target. However, the initial participants to ARET also found that they had divergent perceptions and irreconcilable positions on many issues, especially pertaining to what to do about the substances. Does that mean that something went wrong, that the process was deficient? We claim that a political perspective should complement the rationalistic perspective when considering multistakeholder initiatives. The political perspective recognises that different groups might have opposing views and objectives. Many groups of the civil society, and particularly environmentalists, were hoping that multistakeholder collaboration processes would contribute to the democratisation of decision-making regarding environmental issues. In the specific case of ARET, they were hoping that they would gain a stronger influence on the regulations' formation process. Once they realised that they could not convince the other participants that regulations should be on ARET's table, they decided that they would gain more power outside of ARET and left. In regard of our results showing that current ARET participants are wishing the return of NGOs because they perceive it will improve the legitimacy of the process, leaving the table might turn out to be a negotiation move which will increase the power of NGOs. However, this remains to be seen.

What if there had been a consensus for the mandate of ARET to be a regulatory process as NGOs were asking for? Could a multistakeholder initiative such as ARET reach consensus on such specific issues as regulations? As it was shown in the case, the remaining participants to ARET strongly doubted so. Previous work analysing the substance of consensus reached through multistakeholder collaborative roundtables (MCR) also suggests that consensus on such a specific issue as the detail of a regulation should not be expected as outcomes of MCRs. Indeed, the results of Turcotte (1997) show that, within a MCR, consensus could be reach on general issues and on lists but not on details. So, a first lesson from the case (or a first proposition) is that consensus on specific issues should not be expected from a MCR.

Getz (1995) observed that regulations issued from multilateral agreements were easier to implement when formulated clearly and precisely. So, if consensus on specific issues should not be expected as an outcome of a multistakeholder partnership such as ARET, it does not seem to be an appropriate structure for self-regulation of an industry, which is the claimed objective of voluntary programs. In this regard, the perceived 'lack of leadership' from Environment Canada should be considered not so much as an indicator of the actual performance of this governmental agency but rather as an indicator that multistakeholder initiatives and voluntary programs can not replace regulations. The opponents as well as the promoters of the voluntary approach recognised the need for governmental actions to support the voluntary approach.

It is difficult to create strong incentives for industry action without a supportive legislative and policy context. A recent review of the US federal programs to convince American businesses to reduce pollution voluntarily concluded that most existing initiatives have had minimal benefits either for environmental protection or for industry participants because they were not legislated (Davies and Mazurek, 1996). As Davies and Mazurek conclude in their review of the US initiatives: "Where it regulates, the government will have an obvious lever to backstop voluntary programs. Where it does not regulate, the threat to do so can encourage pre-emptive voluntary action by industry". In short, MCR and voluntary approaches alone should not be expected to allow for self-regulation of the industry in the environmental domain. We can conclude that government role in voluntary initiatives, such as ARET, is essential to earning and retaining public trust and ensuring continued effectiveness.

Is a multistakeholder collaboration process (MCR) a promising method for involving partners? If MCRs should not be expected to replace regulations, however, it could be used in conjunction with regulations or other types of approaches to increase the capabilities for reaching environmental objectives. MCR could be expected to be an agent of transformation of cognitive frameworks (Driscoll, 1995, 1996) and to give direction to actions (Turcotte, 1997), and thus to contribute to environmental governance. From the lessons of the case, we suggest a second proposition: MCR contributes to governance through the transformation of cognitive frameworks and through direction setting (general consensus). ARET provides an illustration of that second proposition. Particularly, the respondents from the industry perceived the establishment of the lists of targeted substances to be a powerful mean of directing their actions.

In the case of ARET, the diversity of the participants was reduced through the process, when the NGOs and the labour groups withdrew. The remaining participants to ARET realised that the legitimacy of the initiative was compromised. This suggests that the diversity of the participants in a MCR is necessary to give credibility to the process. Gray (1985) had proposed that participants must acknowledge the legitimacy of one another for an MCR to be successful. In light of the case of ARET, we further propose that the diversity of the participants is necessary for the process to be legitimate: the diversity of the participants (in terms of the stakeholders they represent) in a multistakeholder partnership is a factor of legitimacy.

REFERENCES

Barouch, Gilles (1989) *La décision en miettes, Systèmes de pensée et d'action à l'oeuvre dans la gestion des milieux naturels*. Paris: Édition l'Harmattan.

Callon, Michel (1993) Comment décider en situation d'ignorance, in *L'État de l'environnement dans le monde* Calliope, Calliope and Bouguerra eds. 294-296. Paris: Éditions La Découverte.

Clarke, Sarah F. (1998) Sustainable Technology: The Role of Networks of Learning, in *Sustainability Strategies for Industry* Roome, N.J. ed. 145-167. Washington: Island Press.

Clarke, Sarah F. and Nigel Roome (1999) *Sustainable Business: Learning-Action Networks as Organizational Assets*. Paper presented at the Academy of Management Conference, Atlanta.

Davies, T. and J. Mazurek (1996) *Industry Incentives for Environmental Improvement: Evaluation of US Federal Initiatives*. Prepared for the Global Environmental Management Initiative, Washington D.C.

Driscoll, Cathy (1995) *Diversity, Dialogue, and Learning: The Case of the Forest Round Table on Sustainable Development*. Ph.D. dissertation, Kingston:Queen's University.

Driscoll, Cathy (1996) Fostering Constructive Conflict Management in a Multistakeholder Context: The Case of the Forest Roundtable on Sustainable Development. *International Journal of Conflict Management*, (April), vol. 7, no. 2:156-172.

Environment Canada (1999) *Consultation Paper on the Future of ARET*. Ottawa.

Environmental Leaders 1,2,3: *Voluntary Action on Toxics Substances*. ARET Secretariat (1994, 1997, 1999). Ottawa.

French, Hilary F. (1995) *Partnership for the Planet: An Environmental Agenda for the United Nations*. Worldwatch Paper 126.

Gallon, Gary (1997) *Voluntary Environmental Measures: The Canadian Experience*. Canadian Institute for Business & the Environment. Montreal, September.

Getz, Kathleen (1995) Implementing Multilateral Regulation: A Preliminary Theory and Illustration. *Business & Society*, Vol. 34, no. 3:280-316.

Gray, Barbara (1985) Conditions facilitating Inter-organizational Collaboration. *Human Relations*, Vol.38, no. 10:911-936.

Healey, P. (1997) *Collaborative Planning: Shaping Places in Fragmented Societies*. MacMillan, Basingstoke.

Hoffman, Andrew J., James J. Gillepsie, Don A. Moore & Wade-Benzoni, A. Kimberley (1999) A Mixed-Motive Perspective on the Economics Versus Environment Debate. *The American Behavioral Scientist*, Vol. 42, no 8:1254-1276.

Kelman, S. (1992) Adversary and cooperationist institutions for conflict resolution in public policymaking. *Journal of Policy Analysis and Management* 11(2):178-206.

Kervin, J. (1992) *Methods for Business research*. New York: Harper Collins Publishers.

Latour, Bruno (1991) *Nous n'avons jamais été modernes – Essai d'anthropologie symétrique*. Paris: La Découverte.

Leiss, William (1996) *Lessons Learned from ARET: A qualitative survey of perceptions of stakeholders*. Working paper Series 96-4. Environmental Policy Unit, School of Policy Studies. Queen's University.

Leiss, William et al. (1997) *Final Report: ARET Voluntary Codes Project*. Prepared for Environmental Affairs Branch, Industry Canada.

Long, Frederick J. and Metthew B. Arnold (1995) *The Power of Environmental Partnerhips*. Orlando, Harcourt Brace & Co.

Miles, M., A.M. Huberman (1994). *An Expanded Sourcebook Qualitative Data Analysis*. Thousand Oaks: Sage.

Minns, David (1997) Voluntary-action' for environmental protection provides business benefits. *Canadian Chemical News*, 8/49.

New Direction Group (1997) Criteria and Principles for the Use of Voluntary Non-Regulatory Initiatives (VNRIs) to Achieve Environmental Policy Objectives, (mimeo)

Porter, Douglas R. and David A. Salvesen (1995) *Collaborative Planning for Wetlands and Wildlife*. Washington: Island Press.

Roome, Nigel J. (1998a) Conclusion: Implications for Management Practice, Education, and Research in *Sustainability Strategies for Industry*. N.J. Roome ed. 259-276 Island Press: Washington.

Roome, Nigel J. (1998b). Introduction: Sustainable Development and the Industrial Firm in *Sustainability Strategies for Industry*. N.J. Roome ed. 1-23 Island Press: Washington.

Strauss, Anselm and Juliet Corbin (1990) *Basics of Qualitative Research. Grounded Theory Procedures and Techniques*. Newbury Park: Sage Publications.

Turcotte, Marie-France (1997) *La prise de décision par consensus: un cas en environnment*. Paris: L'Harmattan.

VanNijnatten, Debora (1998) The day the NGOs walked out: Accelerated Reduction/Elimination of Toxics multistakeholder process' lack of support from non-governmental organisations. *Alternative Journal*, 24/2.

Versteeg, Hajo (1998) *Final Report: Summary of Proceedings of Future of ARET Workshop*. Held December 11, 1997, Ottawa, Ontario.

Yin, R.K. (1994) *Case Study Research, Design and Methods*. 2nd Edition, Thousand Oaks, CA: Sage Publications.

Chapter 14

ESTABLISHING AND MAINTAINING CROSS-SECTORAL ACTOR NETWORKS

Towards Sustainable Development in European Tourism Sector

ZINAIDA FADEEVA

MINNA HALME

1. INTRODUCTION

This study examines a new phenomenon: networking across different public and private sectors of society towards the most important goal of today – sustainable development. The empirical evidence is from the field of tourism, which is among the largest industries of today's world[1]. Travel and tourism is frequently classified as the world's largest and fastest growing industry that provides a contribution into employment, GDP, capital investments, and taxation. Despite the fact that tourism is often offered as a solution to attract development both in the developed and the developing countries, awareness about economic, environmental, and social impacts of tourism activities necessitates the debate about the need for the tourism industry to develop in a sustainable manner.

Tourism stakeholders have recently started to recognise these impacts, but they face problems because single actors, in the field of tourism, are too

[1]The travel and tourism economy contributes 10.8% to GDP and will grow, according to estimates, to 11.6% in 2010. In 2000, the travel and tourism economy accounts for up to 8.0% of total employment. The figure will grow up to 9.1% in the year 2010 (WTTC, 2000). The number of people travelling internationally currently constitutes over 657 million and is expected to rise to about 1.6 billion in the year 2020 (WTO, 2000:21).

T.J.N.M. de Bruijn and A. Tukker (eds.), Partnership and Leadership, 251–271.

small and scattered to take any larger system towards sustainable development. In order to gain momentum to solve various environmental and socio-cultural questions, innovative actors in the field have begun to co-operate in the form of various partnerships like networks. Within such networks, co-operation bridges over business and public sector actors. Compared to manufacturing industries where the relevant actors regarding environmental impacts are often considered to be the ones from along a product chain, within the tourism field also other organisations, for instance municipal agencies are in a crucial role. They set the frame for management of environmental and socio-cultural impacts of tourism. As the 'product' experienced by the tourist is not only the aircraft, accommodation or catering facility, but rather the whole surrounding (city, locality, region, nature area or the like), municipalities actually participate in the larger product creation process. Within tourism, the public-private spheres blend (Welford, Ytterhus and Eligh, 1999).

This study examines joint efforts of networks of multiple actors toward sustainability. By *networks* we understand *co-operative efforts among business firms, governmental bodies or non-governmental organisations, persons, or other entities that are interconnected in various ways* (Smith-Ring, 1997). Since the eight European networks under study seek to further sustainability within the tourism field we will use the term 'sustainability networks'. Despite the term, we do not consider that these networks have reached sustainability. Rather, they are heading toward it. In this article we will discuss composition of sustainability networks, success and failure factors of these networks as well as their contribution to sustainable development.

Even though the context of this study is tourism industry, it may be of interest outside of the tourism context. Apart from a few exceptions (Boons, 1998; Clarke and Roome, 1999) there is little data about sustainable development acted upon in cross-sectoral partnerships. A number of actors in these networks are not solely tourism actors, but public officials or NGOs that represent also other than tourism-related issues. Furthermore, in many focal networks the target of development is region or community. Therefore the study contributes also to a more general body of knowledge regarding cross-sectoral partnerships of multiple actors pursuing sustainable development.

2. DATA AND METHODOLOGY

The study is based on a qualitative analysis of eight tourism networks in five European countries (Box 1). The networks are comprised of actors from tourism business sector, governments and non-governmental organisations (NGO), and academia.

These networks were selected because they had been publicly identified as ones that have taken initiatives towards sustainable development. Some of them were clearly considered pioneers within sustainable tourism. Yet all of them are not 'best practices' but examples of sustainability partnerships in the field. Another selection criterion of the case networks was the *maximising principle* developed by Glaser and Strauss (1967). According to this principle, certain differences between the comparison groups are maximised. In the present sample of networks, differences are, for instance, whether the network is formal or informal, local or dispersed, or whether it has sustainability as its primary or secondary goal.

Box 1. Focal networks of the study

- Alcùdia, Eco-tourism municipality network, Spain
- Calvià, tourism resort's Local Agenda 21 network, Spain
- Finnland Natürlich (FN), farm and nature holidays circle, Finland
- Kinsale, town network, Ireland
- Molyvos, town network, Greece
- Sälen, mountain tourism resort's Local Agenda 21 network, Sweden
- YSMEK, network for developing environmentally friendlier tourism in Finland
- Åre, mountain tourism village network, Sweden

In the local networks the majority of actors is located in the restricted geographical area. They often belong to the same administrative zone or have a common area management and consequently follow the same regulative rules, and often share the same infrastructure. In the dispersed networks, actors are spread over a large geographical area. Some networks are more formal than others in terms of their membership. In YSMEK and FN, for instance, members know that they belong to a network and a list of members is available. On the opposite, the Molivos network does not have a formal membership but the network has emerged from interaction of local actors. In the other networks, there is a formal core or cores (a tight grouping with a lot of interaction between the members and/or clear membership) and in addition a less formal members. For example in the town of Kinsale, the Kinsale Chamber of Tourism (KCoT, the core with clear membership and appointed positions) co-operates with several organisations or individuals of the area. These together form the 'Kinsale network'.

The aim of the study, however, was not the comparison of the networks as such. Rather, the comparison was used for the purposes of finding out what characteristics of sustainability are worthy of further investigation. By comparing the networks to one another, data were generated on practical models and solutions for networking for sustainability, and on the success and failure factors of networks.

The networks were formed at different periods in time and in different institutional settings. Some of them were established quite recently, others have evolved over several decades. The oldest networks of the sample, Molyvos and Kinsale, came through a series of evolutionary changes. Table 1 provides the main characteristics of the networks.

The networks have emerged as response to different types of problems (column *Initial Problem* in Table 1). Kinsale, Molyvos, and Åre areas were confronting the threat to livelihood in the communities. FN was established as a response to the decline of small farms in the Pirkanmaa county in Finland. The network of Calvià, on the other hand, was a response to a threat for carrying capacity. It was formed because the municipality of Calvià experienced a visible decline in the quality of tourism as the destination had degraded. Anticipation of such a threat gave rise to the Alcúdia network. Founders of Sälen and YSMEK networks on the other hand saw an increase in the general awareness in the society concerning environmental questions of tourism enterprises. They decided to apply a network approach to address these concerns.

Surrounding institutions or the lack of those had an impact on the emergence of the networks. The column *Institutional setting vis-à-vis environment* gives a brief picture of the current level of statutory environmental protection in each respective country. This appears to be a significant determinant for a network's aim and operation. For example, the networks of Ireland, Spain and Greece operate in the conditions where environmental infrastructure is either inadequate or not adequately enforced. The main goals of the networks vary considerably from each other (see column *Primary Goal* in the Table 1). Particularly the environmental goals depend to a great extent upon the institutional context. In a context where environmental infrastructure is inadequate, and environmental legislation is not enforced, the sustainability networks tend to prioritise problems like waste management and water emissions. On contrary in countries with adequate environmental infrastructure and legislation enforcement, networks tend to address environmental improvements that go beyond legislation. Finally, it should be noted that the table presents the ultimate goal of the network. The networks also have other objectives that are subordinate to the ultimate goal or instrumental goals for reaching the ultimate goal.

Table 1. Network characteristics

Network	Initial problem	Institutional setting via-à-vis environment	Primary goal
Kinsale (1969) *Formal core & informal members*	Dying town	Incoherent legislation & enforcement, waste & littering are basic problems	Maintaining livelihood of the inhabitants
YSMEK (1993) *Formal*	Environmental effects of hotel and catering industry.	Legislation & enforcement of waste management, waste water treatment. Urban & territorial planning is legislated obligation of municipalities	Voluntary greening of tourism sector in Finland
Finnland Natürlich (1993) *Formal*	Endangered livelihood in farms due to lowering income from agriculture.		Provide an additional source of income to farmers through increase of sales of farm holidays
Molivos (1960) *Informal*	Threat of livelihood, later threat of mass tourism.	Environmental legislation, but deficient enforcement of environmental regulation, waste management is the main problem.	Sustain Molivos in a traditional state and sustain the destination attractive for quality tourists
Sälen (1996) *Formal*	(1) Environmental effects of hotels and (2) construction of slopes and traffic between them.	Legislation and enforcement of waste management, waste water treatment. Urban and territorial planning is legislated obligation of municipalities.	Business: sustain business Municipality: water purification, traffic & road infrastructure
Åre (1994) *Formal cores & informal members*	Threat of losing livelihood in the community.		ÅG/KLG (subnetworks): sustaining business in order to keep community alive VDG (subnetwork): quality of life in the villages
Calvià (1995) *Formal cores & informal members*	Carrying capacity of the destination.	Ecosystems of islands vulnerable and partially damaged. Environmental programs by government, but failed co-ordination or design of environmental legislation and its enforcement	Sustain the destination attractive for quality tourists
Alcùdia (1992) *Formal cores & informal members*	Threat of destination becoming degraded.		Remain as a quality tourist destination.

Explanations of titles of columns:
Initial problem = why was the network originally established. Refers to the past.
Institutional setting = environmental legislation and its enforcement, the level of statutory environmental protection in the country. Refers to the present situation.
Primary goal = the ultimate (main) goal the network aims at (note: the networks have also other goals)

3. NETWORK STRUCTURE

The performance of the networks is defined both by the structures and dynamics established in networks. Ibarra's (1992) analysis of the network studies indicates that most of these studies are concentrated on the effects of the network structures without much attention to the strategies used for enabling and mobilising these structures. Attention to both structural elements and strategies is important due to the mutual dependencies among these elements. Structure affects the dynamic characteristics, while they in turn affect structure (Ibarra, 1992). We will next discuss the structures of the focal networks and thereafter highlight some points regarding dynamic characteristics.

The present networks display a variety of structural forms. All of them are comprised of a mixture of public and private actors. One aspect along which to analyse the network structure is the membership-leadership composition (Table 2).

Table 2. Membership domination and leadership locus

More public actors & public leader	More private sector actors & private sector leader	More private sector actors but public leadership	Mixed
Calvià	Kinsale	Alcùdia	Åre**
Molivos	Finnland Natürlich*	YSMEK	
		Sälen	

* The establisher and financier of FN is Pirkanmaa Agriculture Centre which is a public body, but the co-ordinating office is a private firm paid by the Agriculture Centre.

** In Åre Gruppen (ÅG) and its subgroup Kretsloppsgruppen (KLG) the leadership is held by business actors but the third core of the Åre network, Village Development Group leadership is held by public actors.

3.1 Domain similarity of actors

The structure of networks deserves more scrutiny than only membership domination and leadership locus with regard to public and private sector actors. In addition to government offices and business enterprise actors there are actors from various associations, NGOs and academia in the networks. Moreover, the size and types of tourism enterprises vary across the networks. Likewise the government actors can represent different levels such as local, regional or national government (see Table 3).

Table 3. Diversity of the networks' membership

Net-works	Government similarity		Business similarity***		Associa-tions	NGO	Aca demia
	Horizontal *	**Vertical** **	Enterprise size similarity	Enterprise type similarity			
Kinsale	dissimilar		dissimilar	dissimilar	yes		
YSMEK	dissimilar		dissimilar	dissimilar	yes	yes	yes
FN	similar		similar	moderately similar			
Molyvos	moderately dissimilar		moderately similar	dissimilar	yes	yes	
Sälen	similar		dissimilar	moderately similar			yes
Åre	dissimilar		dissimilar	dissimilar	yes		yes
Calvià	dissimilar	dissimilar	dissimilar	dissimilar	yes	yes	yes
Alcúdia	similar		dissimilar	dissimilar	yes		

* Horizontal similarity of governmental representation
Horizontal similarity = low number of offices/departments represented at one governmental level, e.g. only one municipality office
Horizontal diversity = many offices from the same governmental level in the network. E.g. many ministries.
** Vertical similarity of governmental representation
Dissimilar = offices of government from more than one hierarchical level. E.g. local, regional, national levels
*** Similarity of business enterprises with regard to type or size of member enterprise

In terms of membership base the most diverse network is Calvià. FN and Sälen networks show the highest domain-similarity i.e. their members are the most similar to one another (cf. Human and Provan, 1997)[2]. As such, this tells us nothing. Depending on the network purpose and strategy diverse membership can be a positive or negative factor. It appears on the one hand that diversity of actors provides availability of a high variety perceptions of sustainable development and idea for its implementation. On the other hand high domain-similarity enables fast and efficient implementation of the ideas.

Evidence of both is provided, for example, by YSMEK. The network consists of a management group and pilot enterprises. Management group members are from different ministries, tourism business associations, an environmental group, and an academic. Pilot enterprises enter into different

[2]In FN the majority of members are owners of farm holiday houses and in Sälen large tourism enterprises.

projects (e.g. environmental review, EMS building). They work together with a consultant hired by the management group. Based on the project results YSMEK has published environmental management guidebooks and EMS scheme tailored for tourism enterprises, with the aim of spreading environmental management skills throughout the Finnish tourism enterprises. Diversity of actors in a management group enabled the possibility to assess many views and directions for network actions. At the same time, some of the pilot enterprises acknowledged that the development of conceptual understanding and efficiency of tasks implementation were especially clear within domain-similar groups of companies. Similarly, companies of Sälen network indicate efficiency of implementation in the groups of actors with a similar professional background explaining them by ease of understanding of the problems and realities of the business (professional isomorphism, DiMaggio and Powell, 1991).

3.2 Critical members and powerful outsiders

Our research has shown yet other categories of network actors – critically oriented members and powerful outsiders. Critically oriented members of the networks challenge the network's goals and actions. Calvià and YSMEK have environmental groups as members of the network. Sub-network *Friends of Molyvos* in Molyvos plays a role of 'control factor' in the Molyvos network. This sub-network consisting of intellectuals kept the network under pressure, making municipal decision-makers to resist the mass-tourism mode of local development.

Not all needed stakeholders can become members of a sustainability network. To some extent, effectiveness of network activity will always depend on certain external bodies. For example, Calvià has had a continuous backing from high authorities both at the national and regional levels. High positions of the members of YSMEK management group secured a sufficient level of support for the network activities and for dissemination of the network's results. On the other hand, lack of support from powerful outsiders can hinder network activities. External supporters of KLG (Krettsloppskgruppen = the Ecocycle group), a subnetwork of Åre, were engaged in different interests and did not provide timely assistance for KLG. In Calvià, participation of the construction sector would be critical because securing traditional architecture and halting construction of new large hotels is one of the key cultural sustainability issues. The construction sector, however, has refused to join in due fear of losing control over their activity. This case points out to the need of the special strategy for engaging influential actors whose core activities may be critically affected by network actions.

3.3 Administrative and interactive structures

The above has shown some links between structural characteristics and performance of the networks, but to understand activities and outcomes of sustainability networks, we must also pay attention to certain dynamic characteristics the networks employ to enable their structures. According to Human and Provan (1997) there are two aspects of a network structure – *administrative* and *interactive*. The administrative structure represents a structure with defined roles, responsibilities and rules. The interactive structure is described by more discreet patterns of interaction where the relations between the actors and the prescribed tasks are not clearly pronounced. Both administrative and interactive structures simultaneously exist in a network and serve different functions. The administrative structure is generally established for accomplishing specific objectives. It includes all or any or the following elements: responsibilities, management groups, committees, defined lines of interactions and guidelines for actions. The interactive structures serve a function of delivering consensus, innovations, or understanding in cases of ambiguity.

In many instances, dealing with different aspects of sustainable development involves a high degree of ambiguity. This may be due to the lack of understanding of sustainable development and on the other hand lack of experience about how to implement it. In some instances confusion or uncertainty arises in a situation where a variety of actors are trying to solve a task without much prior experience of working together. Whatever the cause for uncertainty, it clearly calls for a structure that allows development of sufficient shared understanding among actors. Such an understanding is a basis for implementation of ideas.

We will begin by looking at the varied administrative structures of the networks. First, co-ordination in the case networks differed from an appointed management group to an organic group that had over the years assumed the co-ordination of the network (see Table 4). In some networks, the co-ordinators are paid, in others they operate on a voluntary basis (see Table 5).

Table 4. Co-ordination structure

Management group	**Network co-ordinator**	**Organic***
YSMEK	Molivos	Kinsale
Calvià	Sälen	Åre**
Alcùdia	Finnland Natürlich	

* There is not an entity that is called the management group, but rather an entity that developed spontaneously over time and was not appointed in the beginning of the network.

** There is a co-ordinator in the Village Development Group (VDG) and a management in KLG, but there is no formal co-ordination of the whole Åre network.

Table 5. Co-ordination

Paid co-ordinator	Voluntary co-ordination
Finnland Natürlich	Molivos
Sälen	Kinsale
Calvià	Åre*
	Alcùdia
	YSMEK

*VDG's co-ordinator is paid, but KLG is voluntary

Previous research indicates that networks in which the members compete need stronger administrative structure than networks where members have complementary agendas (Human and Provan, 1997). These findings, however, come from business networks and are therefore of limited applicability in the context of cross-sector sustainability networks. In this respect one insight regarding networking of domain-similar enterprises calls for attention. Namely, the network in Sälen consisted of large domain-similar firms (hotels, restaurants and ski lift operators) competing for the same customers. On the contrary, domain-similar actors in Åre network, due to their small size, were able to cover only parts of the market, and consequently were supplying complementary parts of the tourism product. As a result, the Sälen network needed a stronger co-ordination structure to bring together actors whose collaboration was limited due to competition.

To give a couple of examples of the relationship of administrative and interactive structures, the network in Kinsale demonstrates a combination of well-developed interactive structures and relatively relaxed administrative structures based on the voluntary membership in the core group, Kinsale Chamber of Tourism. Within KCoT several committees, among them an Environmental Committee, operate on a continuous basis while other thematic committees can be created on the request of the network members. On the opposite, in FN the administrative structure is dominant. FN is a network that markets its member enterprises holidays to central European, mainly German-speaking market. While joining, the tourism members agree upon achieving a certain quality level of their service. The co-ordinating office is responsible for sales and marketing. The decision for the network's focus and individual tasks is done by the network co-ordinator. The co-ordinator puts quality and environmental management improvements of farm holidays as the prime issues whereas the farm owners and other small tourism enterprise members are more keen on getting their tourism services sold. As the sales has turned out much less than expected, and the enterprise members have not been able to renegotiate the agenda of the network, the activity is stagnating.

Another case of combination of administrative and interactive network structure is demonstrated by YSMEK. The administrative structure of YSMEK is comprised of a management group that decides on the project focus of the networks. The management group hires a consultant, who works directly with the pilot enterprises of the respective project. The pilot enterprises do not have a direct line of communication with the management group in the course of a project nor do they have the possibility to affect the decision of the management group in choosing the network focus. However, the implementation process is acknowledged to be successful. To overcome the fact that the pilot enterprises can join YSMEK only on a project basis, the pilot enterprises established a sub-network, MAYAVAT. MAYAVAT makes it possible to continuously share environmental management ideas and experiences. Others interested enterprises have also joined MAYAVAT. Both the YSMEK management group and MAYAVAT have a creative function of searching new ideas and the ways of implementing them, and they have well-developed interactive structures. Their orientations are clearly different: YSMEK is concerned about environmental questions of tourism at the scale of Finland whereas MAYAVAT essentially deals with practical environmental solutions for member enterprises. They have one joint member who acts a communicator between the groups.

Altogether we found that the interactive structure is more important than the administrative in terms of network outcomes. In other words, the administrative structure cannot compensate the lacking interaction between the members, but it is possible at least to some extent to overcome problems caused by deficient administrative structure if the interactive structure functions well.

4. CONTRIBUTION OF NETWORKS TO SUSTAINABILITY

What then is use of sustainability networks? How can we assess the contribution of networks to sustainable development within tourism? The contribution of a network can be approached from different angles. In this paper we will concentrate on two of them. On one hand, benefits can be distinguished at different areas of sustainability i.e. networks can contribute to environmental, socio-cultural, and economic sustainability (Table 6). On the other hand, we discuss inter-related benefits that sustainability networks can provide to their participants (Table 7). There are a number of actions that simultaneously add value at more than one area of sustainability. For instance, by buying local produce catering and accommodation enterprises of the network contribute to local livelihood and jobs[3]. Simultaneously

[3]By joint purchasing the enterprises of the network manage to create sufficiently substantive demand to support local producers.

environment benefits because the transportation distance of goods is reduced.

Table 6. Examples of benefits from networking at different areas of sustainability

Areas of sustainability	**Contribution at a member enterprise level (examples)**	**Contribution for the community level (examples)**
Economic dimension	*Expected Benefits* Costs savings Image improvement/ Customer satisfaction *Unforeseen benefits:* Generation of new ideas for the business development Improved motivation of personnel	Maintaining livelihood of the community or region through jobs created by tourism
Environmental dimension	Improved waste management Energy savings Value-chain: environmental purchasing, customer education Appointment of environmental co-ordinators	Restoration of natural environmental sites Improvements of local waste separation Water & energy saving plans Traffic reduction schemes
Socio-cultural dimension	Favouring local products Renovation assistance and loans in promoting traditional architecture	Restoration of traditional buildings Full-time instead of seasonal employment Enhancing joint participation in community development & dialogue with local inhabitants

As Table 6 depicts, the actions of networks are either sustainability improvements at member enterprises and activities that contribute to sustainable development of a community. Based on the actions taken by the networks we divided them into three crude categories: (1) community-action networks (ComAct) that seek to enhance sustainable development at a community level, (2) business-action networks (BusAct) that implement sustainability at a business enterprise level and (3) ComBus networks that combine the two approaches. ComAct networks tend to emphasise problem-solution relating to socio-cultural sustainability, such as traditional architecture protection, or the tidiness of a town. BusAct networks, on the other hand, tend to focus on environmental sustainability, mainly environmental management related improvements conducted in tourism enterprises. Calvià, Kinsale, Molivos and Åre form the ComAct group. YSMEK, Sälen, and FN represent the BusAct category. Alcùdia is the only

ComBus network i.e the partnering members clearly try to address both community-level sustainability issues as well as the enterprise sector's sustainability issues[4]. This is not to say that ComAct networks would not engage in any action regarding the enterprise sector or that BusAct networks were totally restricted to working with sustainable development of enterprise-level action. The categorisation indicates that the majority of the network's activities lie at the respective level, enterprise or community.

One would expect that ComAct networks are those that are lead or dominated by public sector members. Vice versa, BusAct networks should be lead by or they should primarily consist of tourism enterprises. Unexpectedly, however, ComAct networks do not consist predominantly of public sector actors. Despite their community-level orientation of sustainability efforts, ComAct networks can be business-dominated in terms of the number of members and steering responsibility. Conversely BusAct networks did not consist principally of business actors. Some BusAct networks are being steered by public sector actors, for example, municipality authorities or ministry representatives. This implies that the membership composition does not determine the activities a network chooses on its agenda.

In addition to the contribution vis-à-vis sustainable development, we can identify benefits that a network brings about to a person, firm, industry and society (local community or region). From the motivational viewpoint the significance of the latter set of benefits should not be underestimated. Table 7 depicts these additional benefits of sustainability networking. Some of the items are overlapping, i.e. they could be categorised at more than one level. However, for the purpose of clarity we have categorised each benefit to a level which seems either the most obvious or where its biggest impact lies.

[4]For example, Alcùdia has recovered monuments of historical interest, and renovated traditional buildings. On the other hand, the Alcùdia network has an environmental label which requires among other things that hotels and restaurants conduct energy and water saving measures and serve traditional local dishes.

Table 7. Benefits from networking experienced at different levels

Benefiting level	Value-added
Person-level	Improvement of working conditions Ability to live according to one's values A feeling of 'belonging to something' (to a network) Civic pride of the region or community Anti-bureaucratic sentiment Empowerment through self-help Ensured livelihood for the present and future generation
Firm-level	Information increase for the firm belonging to the network Image improvement and competitive advantage Resource optimisation and/or cost savings
Industry-level	Networks serve as examples of good practices Change of the status of the industry
Society-level	Increase of the attractiveness of the region through networking Increase of the region's lobbying power

5. ASSESSMENT OF THE NETWORK RESULTS

How significant are the contributions of the studied networks with regard to sustainable development? The response to this question calls for consideration of at least two things. First, we should consider what are the most pressing sustainability problems in the area of network influence. Second, we need to assess the achievements against the goals a network sets for itself. The first point calls for some external evaluation of the network context whereas the second point is based on the accounts of the network members. Table 8 presents the main sustainability-related problem of each network, their primary goal (as seen by the network actors), the main outcome of the network in a broad sense, and assessment of the outcome. Evidently the table is crude simplification of reality. It appears that in a broad sense the Kinsale, Calvià and YSMEK networks have reached their main goal. The others show some deficiencies in this respect, FN and Sälen the most. The following discussion about success and failure factors will highlight some critical issues related to network outcomes.

Table 8. Network characteristics

Network (established in)	**Initial problem**	**Primary goal**	**Main outcome**	**Assessment/ deficiencies**
Kinsale (1969)	Dying town	Maintaining livelihood	Lively tourism destination	Goal reached, but tourism development threatens SD
YSMEK (1993)	Environmental effects of hotel & catering industry	Voluntary greening of tourism sector in Finland	Impact given to greening of tourism enterprises	Own goal reached. Little concern of socio-cultural SD
Finnland Natürlich (1993)	Endangered livelihood in farms due to lowering income from agriculture	Endangered livelihood in farms due to lowering income from agriculture	Sales not increased. Some environmental. & cultural heritage improvements among active farms	Own goal not reached. Less environmental improvements than expected
Molivos (1960)	Threat of livelihood, later threat of mass tourism	Sustain Molivos in a traditional state in order to sustain the destination attractive for quality tourists	Keeping mass tourism out	Own goal reached vis-à-vis socio-cultural SD, but environmental problems remain
Sälen (1996)	(1)Environmental effects of hotels & (2) construction of slopes & traffic between them	Business: sustain business Municipality: water purification, traffic & road infrastr.	Laying ground for future environmental actions. Halting expansion of area.	Own goals not reached. No major contribution to SD
Åre (1994)	Livelihood in the community	ÅG/KLG: sustaining business in order to keep community alive VDG: quality of life in the villages	Paving the way for future sustainable life in Åre	Too early to assess. Some environmental & social actions done, but continuation is not sure
Calvià (1995)	Carrying capacity of the destination	Sustain the destination attractive for quality tourists	Groundwork for SD done well, some waste collection & water saving activities conducted	Own goals reached.
Alcùdia (1992)	Threat of destination becoming degraded	Remain as a quality tourist destination	Environmental improvement of hotel sector, development of cultural tourism	Own goals to some extent reached. Municipal waste mgt system remains a problem.

6. SUCCESS AND FAILURE FACTORS

Many factors can be responsible for the success or failure of complex undertakings of inter-organisational networks. These factors may be at micro- or macro-level or relate to contextual issues. In some cases the success of an initiative may depend on somebody's leadership, the general level of awareness of the participants, the common background or previous working culture of the network actors. In other instances regulatory barriers or lack of resources may hinder it. In this section we do not attempt to provide a comprehensive overview of all the factors that are critical in a sustainability networking process. Instead we will discuss the most striking examples of critical factors that appeared to facilitate or hinder the networking process during network establishment and network maintenance.

6.1 Establishing a sustainability network: critical factors

Clarity of goals appeared a significant success factor. In some networks unclear or contradictory goals lead to the dysfunction of the entire network while in others it only partially hindered activities. In the above mentioned example of FN, for instance, the co-ordinator considered quality and environmental improvements as the primary goal of the network whereas the farm holiday houses and other members were more keen on increasing sales. The lack of goal clarity often resulted in *un-met expectations*. In Alcùdia, the municipality initiated a Local Agenda 21 process, which lead to the network formation. The Tourism Board that mainly consists of business enterprises was keen on developing the town in environmental and cultural aspects together with the municipality offices. However, they became disappointed with the purely informative meetings run by the Town Council. They felt that the municipality only expected enterprises to engage in concrete actions but did not fulfil their expectations like improvements of the municipal waste management. The confusion about network goals was mainly caused by a lack of communication in the establishing stage.

It may well be that in networks in which actors represent quite different perceptions about how implement sustainable development, totally unified goals are an unattainable ideal (Boons, 1998). It appeared that to a great extent the problem of unclear goals was due to lacking communication. To overcome this kind of problems, the network members ought to discuss about their expectations and objectives already at an early stage, and continue the discussion as the co-operation continues. This provides the opportunity to elaborate goals and strategies. All to often in our case networks there was predominantly one-way communication, flowing from leader or 'hub actor' to other members (cf. Larsson et al., 1998). This meant that feedback and ideas from the 'recipient' members were not utilised.

Moreover, their disappointment about unmet expectations did not come into the awareness of the leading actor.

Determination of the optimal membership is another critical factor. In multi-stakeholder public-private networks the actors vary considerably regarding their knowledge bases and dominant logic of action. On the one hand, there need to be different types of members to make a broader understanding of sustainable development available. On the other hand, the actors of a network need to be able to create a certain amount of common ground in order to act. For example, the diversity of management group of YSMEK lead to the generation of interesting and well supported projects, while the similarity of the companies participating in pilot projects resulted in a fast and efficient implementation process.

6.2 Maintaining a sustainability network: critical factors

Dealing with multiple actors and complex problems requires the establishment of the formal structures that would allow 'getting things done'. At the same time too much administration seemed to be detrimental for the creativity of the members. Adequate *combination of flexibility and formality* appeared to serve several functions. It allows enthusiastic actors to take responsibilities within the network irrespectively to the positions they happen to otherwise occupy. It also allows the emergence of new sub-networks that usually provide an additional information sharing media and may speed the implementation of certain actions.

The issue of flexibility and formality closely relates to another important structural factor – *feedback for adjustment of the network strategy*. The networks have shown the need for the mechanism that would allow modification of its structure and strategy to the changing situation outside and evolving experiences inside the network.

The studied networks were affected positively by the early experiences of that could be called '*small wins*'. Small wins are controllable opportunities of a modest size that produce visible and tangible outcomes (Weick and Westley, 1996). They reinforce the motivation of actors to keep up with the activity, and they also serve as a source of learning. Sometimes the question of creating small wins is simply a matter of articulation of the network's achievement. In other words, the actors may not even realise what has been accomplished. Despite the importance of small wins, it is advisable to adopt a *long-term time orientation* when starting a network. The idea of early tangible gains does not remove the fact that as network members share their context over time, they should become more adept at explicating tacit and embedded knowledge to one another, and develop the network fit for creating 'large wins' (Halme, 2001).

6.3 Trust

Trust is one of the key factors influencing network success (Carney, 1998). We found that in essence trust manifests itself in various ways in different stages of the network evolvement process. One of the conditions for the network success is an *initial level of trust* among the actors. For example, the business actors of the Sälen network were long-standing competitors. Co-operation in the Local Agenda 21 was a novel experience. Missing collaborative experience and lack of initial trust made the actors act in a very consciously calculative way toward each other. In contrast, the actors of the Åre network have complementary services within the same geographical area. There was a high degree of initial trust based on the long and non-conflicting history of co-operation.

Credibility of network actions contributes positively to trust building. The network in Calvià took an effort to obtain credibility through series of actions like: halting of construction projects; demolition of twelve degraded hotels in order to create green areas on the seaside; transformation of roads into a boulevard and contracting the ecologist group GOB for the environmental education programmes in schools. These activities served as a source of credibility of the network's intentions among its members (and also to outsiders).

The norm of *reciprocity* influences the trust building process. Reciprocity implies that in multi-actor co-operations, only those actors willing to give something will receive something in return (Lütz, 1999; Sydow and Windeler, 1998). In FN, the enterprise members expected the office to provide them with more customers, and except for a few active members, they themselves were not willing to invest in the network. In Sälen, the enterprise actors wanted the local agenda 21 co-ordinator to run the network, and did not spend efforts to keep the network action going when the co-ordinator changed. In Alcùdia, the enterprise members felt that the municipality requires improvements from them, but does not engage in similar improvements itself. In other words, the expectations of 'getting' were higher than the willingness to 'give' in these networks. In Sälen, the rule of reciprocity appeared to have been violated before the local agenda 21 network was started. This negatively influenced the network co-operation from the beginning. Historically, there was an adversarial relationship between the tourism businesses and municipality in the area. The tourism business representatives claimed a failure of the municipality to respond to their demands with respect to e.g. adequate provision of social and municipal services. This has constantly undermined trust. Business enterprises set the condition for the municipality to take equal financial obligations in the future activities. Instead of engaging in the LA21 process properly, they expect the municipality to take the responsibility, and to 'show' that this time the municipality is really going invest in the work.

7. CONCLUSIONS

The focal public-private networks display a variety of structural forms and dynamic characteristics. It appears that the membership composition does not determine the network's actions. Networks, where the majority of members come from the private sector can concentrate their actions both on the level of business enterprises as well as on the community level. The same holds true for the networks comprised of a majority of public actors.

Although the leadership locus (i.e. private or public leadership) varied among the networks, one observation remains unchanged: too strong a *leadership* may destroy the activity of other network members. While co-ordination and leadership secure efficient implementation of network actions and provide additional information sharing media, they may suppress the creation of new ideas, natural re-formation of the network, and the emergence of the subnetworks.

While an *administrative structure* secures accomplishment of the specific objectives, it cannot compensate for the absence of the *interactive structure*. Interactive structure proved to be essential in dealing with complex and ambiguous tasks dealing with sustainable development. It was especially important when actors lacked a prior experience of co-operation.

Tourism actors – like most actors in society – have a fragmented and occasionally contradictory understanding about environmental, socio-economic and cultural aspects of sustainable development, and they implement it depending on pressing problems and their awareness level. A too-narrow membership basis may limit the understanding of sustainability within the network, whereas an overly large number of members may hinder implementation. When establishing a network, a crucial success factor for networking among different public or private actors is the development of a *common vision*. This process of articulating individual values and preferences should lead to the identification of a sense of commonly accepted goals. The *membership composition* of a network has to be relative to the goal. A network that include varied expertise of the respective area of sustainable development and also actors that are influential with regard to the specific sustainability goal proved to be more effective.

Most of the present policy measures assist in *establishing* sustainability networks. Attention should also focus on *maintaining* their viability and effectiveness in the course of time. 'Small wins' created fairly early to support the network's learning for sustainable development, affect the actors' motivation and provide possibilities for network adjustment. However, it is advisable to adopt a long-term orientation when starting a network. As trust between diverse actors is based upon previous interactions, a network needs time to become adept at explicating tacit knowledge among its actors, and to develop the network so it can facilitate the creation of more fundamental results. *Long-term orientation* also applies to *public funding providers*: networks should not be expected to run on their own after two or three years. Short-term funding has a tendency to lead to fragmentary results and to thwart the possibilities of acting upon sustainable development, at a local level.

The motives of network actors for working towards sustainable tourism are usually interrelated with other motives, experienced at a *personal* level, for the *firm* or for the *municipality/region*. For instance, improvement of the livelihood of a region or competitive advantage of an enterprise is such a motive. These are important from a motivational viewpoint, especially if networking is on a totally voluntary basis.

ACKNOWLEDGEMENTS

This project was conducted within the Environment and Climate Programme 1994-1998 (FP4) of the European Union. We would like to thank Silvia Ayuso, Ger Mullally and Maria Hatziantoniuou from the EMPOST-NET team for their contribution to the project.

REFERENCES

Boons, F. (1998) Caught in the Web: The Dual Nature of Networks and Its Consequences. *Business Strategy and the Environment* 7(4):204-212.

Carney, M. (1998) The Competitiveness of Networked Production: The Role of Trust and Asset Specificity. *Journal of Management Studies* 35(4):457-479.

Clarke, S., N. Roome (1999) Sustainable Business: Learning-Action Networks as Organisational Assets. *Business Strategy and Environment* 8(5):296-310.

DiMaggio, P., and W. Powell (1991) The Iron Cage Revisited: Institutional Isomorphism and Collective Rationality in Organisational Field in *The New Institutionalism in Organizational Analysis,* W. Powell and P. DiMaggio eds. 267-292 Chicago: University of Chicago Press.

Glaser, B. and A. Strauss (1967) *The Discovery of Grounded Theory: Strategies for Qualitative Research.* Hawthorne, NY: Aldine Publishing Company.

Halme, M. (2001) Learning for Sustainable Development in Tourism Networks. *Business Strategy and the Environment,* 10 (2):100-114.

Human, S. and K. Provan (1997) An Emergent Theory of Structure and Outcomes in Small-firm Strategic Manufacturing Networks. *The Academy of Management Journal,* 40(2):368-403.

Ibarra, H. (1992) Structural Alignments, Individual Strategies, and Managerial Action: Elements Toward a Network Theory of Getting Things Done. in *Networks and Organizations: Structure, Form, and Action*. N. Nohria and R. Eccles eds.

Larsson, R., L. Bengtsson, K. Henriksson and J. Sparks (1998) The Interorganizational learning Dilemma: Collective Knowledge Development in Strategic Alliances. *Organization Science* 9(3):285-305.

Lütz, S. (1999) Learning through Intermediadies: The Case of Inter-Firm Research Collaborations. in *The Formation of Inter-Organizational Networks*. M. Ebers ed. 220-237 Oxford: Oxford University Press.

Smith-Ring, P. (1997) Processes Facilitating Reliance on Trust in Interorganizational Networks. in *The Formation of Inter-Organizational Networks*. M. Ebers ed. Oxford: Oxford University Press.

Sydow, J. and A. Windeler (1998) Organizing and Evaluating Inter-Firm Networks: A Structurationist Perspective on Network Processes and Effectiveness. *Organization Science* 9(3):265-284.

Weick, K. and F. Westley (1996) Organizational Learning: Affirming an Oxymoron. in *Handbook of Organization Studies*. S. Clegg, C. Hardy, W. Nord. London: Sage.

Welford, R., B. Ytterhus and J. Eligh (1999) Tourism and sustainable development: An analysis of policy and guidelines for managing provision and consumption. *Sustainable Development*, 7:165-177.

WTO (2000) *Tourism Highlights*.

WTTC (2000) *Tourism Satellite Accounting Research, Estimates, and Forecasts for Governments and Industry*.

Chapter 15

Stakeholder Involvement and Alliances for Sustainable Households

The Case of Shopping, Cooking and Eating

JACO QUIST

KEN GREEN

KLÁRA SZITA TÓTH

WILLIAM YOUNG

1. INTRODUCTION

Achieving sustainability requires profound changes in both our consumption patterns and present methods of production. Making production and consumption systems sustainable, however, implies the involvement and support of many stakeholders. Today, it is widely supported that changes necessary on the level of systems require the involvement of all the main societal groups including governments, companies, public interest groups, the public and research bodies (e.g. Vergragt and Jansen, 1993; Weaver et. al., 2000). However, involvement alone is not sufficient as each societal group should actively contribute to the necessary changes within their specific roles and possibilities. Clearly, one societal group or a single stakeholder cannot carry out all necessary activities, as it commonly requires co-ordination and co-operation among and between social actors and stakeholders. Co-operation can take the form of concrete partnerships and alliances, while for instance, Roome (2001) emphasises the role of environmental networks of stakeholders that work together based on a shared vision.

This chapter takes as a starting point that achieving sustainability requires factor 10-20 environmental improvements in production and consumption

T.J.N.M. de Bruijn and A. Tukker (eds.), Partnership and Leadership, 273–293.

systems in the long term. Factor 20 is derived from the so called IPAT[1] formula (Ehrlich and Holdren, 1974) and is based on a doubling of the world population combined with a five-fold increase of wealth per capita while halving the total global environmental burden. Such environmental improvements require combinations of technological, cultural, organisational and institutional-regulatory innovations - for which we use the term system innovations in this chapter -, and assume strong stakeholder co-operation and contributions from all societal groups mentioned earlier. We also argue that bringing about factor 20 system innovations requires new integrated approaches that should combine:

- The involvement of a broad range of stakeholders and actors from different societal groups not only when defining the problem but also when searching for solutions and conditions and developing shared visions.
- Incorporating not only the environmental component of sustainability, but also its economic and social component.
- Taking into account both the demand side, consumption and the supply chain focussing on production and consumption systems.

Within this chapter, we describe part of the EU funded research project 'Strategies towards the Sustainable Household' (SusHouse)[2] in which such an integrated approach using stakeholder workshops and participatory scenario construction was developed and tested in five European countries. This approach is illustrated using results from three "Shopping, Cooking and Eating" (SCE) case studies[3] and we focus on how we involved stakeholders in processes of constructing normative scenarios depicting sustainable futures, how involved stakeholders contributed to defining follow-up activities and implementation strategies, and what kind of stakeholder co-

[1]The presently popular version of the IPAT equation is I=PxAxT: the environmental impact (I) equals the product of population size (P), the degree of affluence (A) per person, and the environmental impact from technology (T) used to produce one unit of affluence.

[2]The SusHouse project was a collective endeavour of six research groups in five European countries, these were: Technology Assessment Group, Delft University of Technology, (the Netherlands); Szeged College of Food Industry (Hungary); Dept. of Industrial Design, Politecnico di Milano (Italy), Avanzi (Milano), Manchester School of Management, UMIST (UK), Lehrstuhl Markt und Konsum (Universitat Hannover, Germany). The contributions of all involved researchers to the project methodology and discussions have been essential to the work described in this paper. This research has been supported by the EU DG 12 Environment and Climate RTD Programme, Contract no. ENV4-CT97-446.

[3]For an overview of the full results of these three case studies carried out in Hungary, the Netherlands and UK, see Green and Young 2000, Quist 2000, Toth et.al. 2000, Young 2000. More information about other elements and cases can be found on the website www.sushouse.tudelft.nl, or in the final summarising report (Vergragt, 2000).

operation around concrete follow-up activities was realised. Therefore, this chapter deals with questions like:

- How did the involved stakeholders contribute to designing the sustainable futures (through normative scenarios) which were addressed as possible sustainable alternatives for particular lifestyles?
- How to identify and enrol stakeholders that are essential for realising these possible sustainable futures depicted by the scenarios?
- What kind of stakeholder co-operation is necessary for realising the next steps towards each scenario?

Concerning the outline of this chapter, it starts by briefly discussing the methodology of the SusHouse project and the SCE function (section 2). Then it describes the process of stakeholder involvement and the results of the case studies (section 3), including an evaluation of the state of implementation, follow-up and emerging stakeholder co-operation. Some conclusions are drawn in section 4 in which we also address the extent to which partnerships between stakeholders were created, the elements in the SusHouse approach that were essential and the possible positive and negative factors.

2. SUSHOUSE BACKGROUND AND SHOPPING, COOKING AND EATING CASES

2.1 SusHouse background

The SusHouse project was concerned with developing and evaluating strategies for transitions to sustainable households (Vergragt, 1999; Vergragt, 2000). The project took as a starting point that achieving sustainability requires factor 10-20 environmental improvements in production and consumption systems over the next 50 years. The overall aim was to develop and test a methodology enabling companies, governmental organisations and NGOs to carry out their own analyses of sustainable function fulfilment. Further objectives included the development of normative scenarios for the sustainable fulfilment of household functions endorsed by a range of social partners, scenario assessments and the construction of follow-up agendas containing implementation proposals, stakeholder co-operation around concrete proposals, and the development of policy recommendations and further research questions (Vergragt, 1999, 2000).

The SusHouse project involved six research groups from different disciplines in five European countries, while three household areas where studied: Clothing Care, Shelter and Shopping, Cooking and Eating. Both consumption and its interconnection with production through products and product usage were taken into account, while the involvement of stakeholders in the process of (re)designing the fulfilment of a household's needs through stakeholder identification and workshops was one of the main elements. The approach was applied earlier in the Sustainable Washing project at the STD programme (Vergragt and Van der Wel, 1998), but it was elaborated and extended further during the SusHouse project. Major changes included (Green and Vergragt, 2001):

- Less emphasis was laid upon technology as the main agent for sustainable development; rather, a combination of *technological, social, and cultural changes* was envisaged.
- More emphasis was laid upon the *participation of non-governmental stakeholders* in the process.
- A *design orientation* was chosen, rather than a policy-making orientation.

The applied approach is briefly as follows, focusing here on the workshop organisation and stakeholder identification methodology[4]. For each household function studied in each country, an extensive process of *stakeholder identification* was performed, covering stakeholders on the demand side, the supply side, research bodies, government and public interest groups. Selected stakeholders from different stakeholder groups attended stakeholder *creativity workshops* aiming at identifying sustainable ways of future function fulfilment and generating new ideas to achieve this. The results were used by the research teams for *scenario construction.* These normative scenarios were not only assessed in terms of environmental gain, consumer acceptance and economic credibility, (this part of the SusHouse approach is not dealt with in this chapter), but also for a second round of scenario-specific stakeholder identification. Old and newly identified stakeholders were invited to a second set of *strategy and implementation workshops* in which scenarios and assessment results were discussed, followed by developing implementation proposals, research agendas and policy recommendations for achieving the scenarios. In both series of workshops *backcasting techniques* were applied. Backcasting means looking back from a desired future (in our studies depicted in normative scenarios) in order to identify how we could have got at the wished future before defining first steps and activities towards that future.

[4]A more extensive elaboration on the overall SusHouse methodology is given by Vergragt (1999, 2000), Quist et al. (2001) and with respect to the workshop organisation and stakeholder management methodology by Quist, Pacchi and Van der Wel (2000).

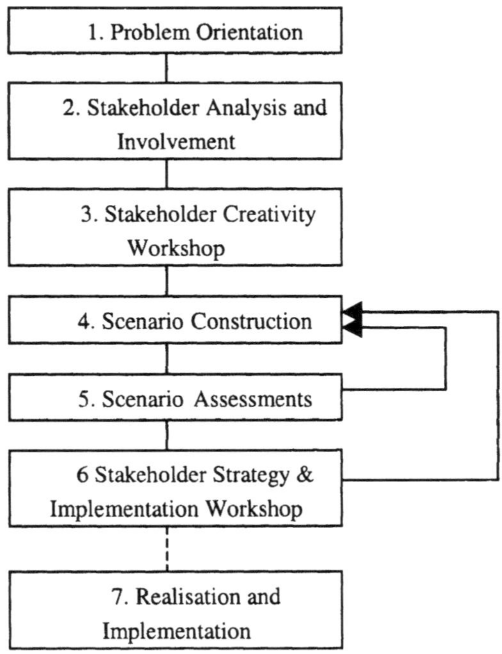

Figure 1. The stages in the SusHouse project

Figure 1 shows the six stages of the overall SusHouse approach including possible iterations which can be applied to improve the normative scenarios. The seventh stage concerning 'realisation and implementation' was actually not part of the project. However, an important aim of the project was to stimulate and generate follow-up based on the outcomes of the case studies.

2.2 Shopping, Cooking and Eating case studies

In the SusHouse project the Shopping, Cooking and Eating (SCE) function was studied in Hungary, the Netherlands and the United Kingdom. It is a very complex one covering agriculture, and the inputs to agriculture, through food processing, shopping, home storing, cooking, and eating, to the disposal of waste and consumer durables. The food production system includes food industry, primary agricultural production and their suppliers, and is a major employer in all countries considerably contributing to the economy, but also a huge user of energy and other inputs (Tansey and Worsley, 1995). It involves issues of nutritional balance and adequacy, and incorporates strong cultural and emotional aspects. It also includes decisions of household members influencing consumption and purchasing patterns, determining (Green and Young, 2000; Quist et al., 1998):

- the balance between eating-in and eating-out/home delivery/take-away;
- the timing and types of meals eaten;
- the balance between pre-prepared and unprepared food;
- the organisation of cooking and shopping; and,
- how much food is grown in hobby gardens and how much is bought.

Obviously, decisions of household members have direct implications for the food service sector (eating-out versus take-away and home delivery); the retailing sector (purchase from supermarkets or smaller food retailers, direct purchase from wholesalers, direct purchase from growers); and, the food-processing sector as different patterns of household eating will change the balance between different processing methods (Green and Young, 2000; Quist et al., 1998).

Table 1. The four level food production and consumption system as defined for SCE

Level 1: The "core" function Shopping, Cooking, Eating The core covers all activities of the household directly associated with the purchase, storage, cooking, eating and clearing up of food and drink, this includes: – 'Home' cooking, storage, clearing up after eating (including washing-up) – Purchase from take-aways for eating in streets or at home, including deliveries of cooked food and meals – Eating in restaurants and work canteens or office canteens – Production of food in householders gardens or small-scale plots – Household transport associated with the purchase of food from shops and supermarkets, take-ways and transport to restaurants
Level 2: Distribution All retailing and wholesale of food, including the transport associated with this
Level 3: Agriculture, Processing and Durables – Production of basic food products (i.e. agriculture, including imports) – Food processing of every kind – Manufacture of kitchen equipment (for storage, food preparation, cooking, eating and clearing up) – All transport associated with this, moving foods from farms etc. to factories and between factories
Level 4: Delivery of inputs into other levels Manufacture of inputs into other levels, including agricultural chemicals, fertilisers and packaging

Based on all these assumptions a general four level food system was defined from the households' point of view (see Table 1) emphasising the core level of shopping, cooking and eating, whilst also including storage (Quist et al., 1998; Green and Young, 2000). This description is in general terms, as the precise structure of each level will differ for the three countries (and also within the three countries). Table 2 illustrates some differences by highlighting some current characteristics of shopping, cooking and eating in households in the three countries (Green et al., 1999).

Table 2. Current SCE characteristics in Hungary, The Netherlands and the UK

	Hungary	**The Netherlands**	**United Kingdom**
Population (in millions)	10.1	15.5	58.6
Number of households in millions (and averaged number of household members)	4.0 (2.5)	6.5 (2.4)	23 (2.5)
Average expenditure per household on food per year	1,028 Euro (33.1%)	3,640 Euro (19.2%)	3,230 Euro (19.4%)
Proportion on food shopping	96%	80%	72%
Proportion on eating out	3%	20%	18%
Proportion on take-away	1%	Unknown	10%

2.3 Trends and un-sustainabilities

After defining the system, it was possible to identify its unsustainabilities which was done in an early stage of the project with the help of desk research and expert interviews. Results from our early explorations showed that unsustainabilities were very similar in UK and the Netherlands, while it was partially different in Hungary (which also highlighted environmental problems like water quality, sewage problems and air pollution). Results are elaborated elsewhere (Quist et al., 1998), and only briefly summarised here.

Concerning the demand side, in UK and the Netherlands, there is a growing consumption of processed foods, high energy varieties and exotic vegetables. Those are either grown in fuel driven greenhouses or are to a growing extent imported by aircraft. Furthermore, eating-out is growing steadily, while its energy requirement is much larger than for meals prepared at home. In all three countries there is an increase in the SCE related energy usage in households. In the Netherlands this is partly due to an increasing volume of kitchen equipment, while in Hungary it was also due to low energy efficiency of old equipment. Growing volumes of waste and packaging (both in-house and take-away) were mentioned in each country. Wasting food and human behaviour were emphasised in UK and the Netherlands.

Concerning agriculture, the main unsustainabilities include intensive livestock breeding (energy usage, eutrophication, acidification), pesticide and fertiliser usage in arable land farming (energy, ecotoxicity) and intensive greenhouse horticulture (fossil fuels and pesticides). It was also found that stakeholders and experts can have different opinions on the unsustainabilities of the food system. For example, packaging was considered an important unsustainability by environmental groups in the Netherlands, while food experts and food processors argued that in the Netherlands packaging had only minor impacts compared with the environmental impact of the foods due to efficient waste processing (re-using heat) and composting biological waste. The most likely explanation can be found in the different mental reference frames and values of, in this case, grass root groups and companies.

In order to achieve some idea of future function fulfilment in the early stage of the SCE case studies, consumer trends and other developments in the food system were also identified through desk research and stakeholder consultation (Quist et al., 1998). General consumer trends and demographic trends include ageing, sub-urbanisation, mass individualisation, a growing number of singles, growing health concerns (e.g. BSE), growing demand of convenience and variety, growing multi-ethnic influences, growing concern of sustainability, animal welfare and community affairs, and a growing demand for quality. In Hungary, the poor wealth growth was of concern too, while hobby gardens are expected to remain important for local food supply for the next decades. Related SCE trends include an increase of grazing, eating-out, home delivery, ready-mades and ready-made meals, and also faster cooking through assembling meals using meal components and intelligent kitchen equipment. In all three countries the share of supermarkets and hypermarkets is growing. Identified supply chain developments included chain reversal (becoming more demand oriented), consumer driven food technology development, development of new intelligent kitchen equipment, increase of organic growing and gradual changes in mainstream agriculture and livestock breeding. This exploration also revealed that Hungary is actually having similar trends and developments as UK and the Netherlands, though these developments are in an 'earlier' stage.

2.4 Stakeholders and future stakeholders

It has already been mentioned that stakeholder participation was a crucial element in the SusHouse approach. Stakeholders were defined as all those individuals and organisations having an interest at stake, or those that can affect SCE or can be affected by changes in SCE. Stakeholder participation was among other reasons important for getting access to their knowledge and getting familiar with their (value) driven opinions, while involvement is also crucial when aiming at implementation and follow-up.

The key assumption behind this approach is that interaction and exchange of knowledge, values and opinions between stakeholders contribute to so-called social learning (both on the level of facts and cognitive knowledge and also on the level of mental frameworks, values and value driven opinions), while it also enables the 'social shaping' of future visions of need fulfilment and technology development (Quist, Pacchi and Van der Wel, 2000; Quist et al., 2001). Hence, the SCE case studies can also be described as interaction processes in which stakeholders learn from each other and on sustainable future options also affecting their frameworks and paradigms. It may, however, also limit the scope of new creative solutions that are too disruptive when comparing it to present situation. It will be clear that an important question was how to identify and select the right stakeholders in several stages of the project including future stakeholders that do not have a stake in the present system, while how to stimulate learning on all levels in order to broaden support and acceptance for solutions and new ideas was important too.

Table 3 gives an overview of possible stakeholders and stakeholder groups as defined in the early stages of the project. Stakeholder categories were distinguished in all levels of the defined system, and grouped into four societal groups (while adding an additional fifth group):

1. The *public sector* including public interest groups and consumers;
2. *Companies*, their trade organisations and other representatives;
3. *Governments* including government councils and government related organisations;
4. *Research bodies* including research institutes and universities; and
5. *Others*, including media, opinion leaders and educational bodies.

Obviously, Table 3 contains rather general descriptions of relevant stakeholder groups within the five categories. However, within the stakeholder identification and involvement activities in each country organisations, companies and appropriate contact persons were identified.

Concerning the issue of future stakeholders, this was mainly done in the second scenario specific round of stakeholder identification.

Table 3. Stakeholder groups in the food system of Hungary, The Netherlands and UK

Level	Stakeholder category	Societal groups
1	Consumers and household members	Public
	Consumer organisations	Public
	Food service suppliers (restaurants, fast food chains, take-aways, home deliverers, canteens, special caterers)	Companies
2	Supermarkets (big and small)	Companies
	Specialised food shops (baker, butcher, fish seller etc.)	Companies
	Distributors, wholesalers	Companies
	Kitchen equipment sellers	Companies
3	Kitchen apparatus manufacturers	Companies
	Food processors	Companies
	Horticulturists, their organisations and vegetable/fruit processors	Companies
	Plant growers, their organisations and crop processors	Companies
	Animal breeders, their organisations and meat processing industry	Companies
	Fishers and fish processors	Companies
	Vegetarian protein food producers	Companies
	Organic growers and breeders and their organisations	Companies
4	Domestic and industrial energy suppliers	Companies
	Domestic and industrial water suppliers	Companies
	Packaging producers	Companies
	Manufacturers of inputs to level 3 (fertiliser, chemicals, animal feed stuffs etc.)	Companies
Other	Specific units of ministries	Government
	Environmental movement	Public
	Animal welfare organisations	Public
	Specific research groups at universities and research bodies	Public
	Opinion leaders, consultants	Other
	Labour unions	Public
	Media	Other
	Relevant research institutes, universities and related bodies	Research

3. CASE STUDY RESULTS

3.1 Stakeholder involvement and process results

During the case studies two rounds of stakeholder analysis were performed before each round of workshops. The first round focused on present stakeholders, their stakes and the global objectives driving them in the SCE field. The second round existed of a general update of the first round, while it was extended with a scenario specific part identifying driving forces in the constructed scenarios and imagine how these could affect present stakeholders and how this could lead to the emergence of new stakeholders in the field. Though these activities led to balanced set of stakeholders in each country, it does not reveal the stakeholder involvement results. This was done by analysing the participants' numbers and backgrounds at the workshops. Table 4 contains the results and shows attendance numbers for each of the five main stakeholder groups distinguished earlier in this chapter. Figures are given for distinguished stakeholder categories with respect to participation in the two workshops and the stakeholder and expert interviews taking place before the first workshop.

Table 4 shows not only that recruitment and the enrolment of stakeholders in both series of workshops was successful, but also that the distribution over the five distinguished societal groups was in general quite balanced. Additionally, in Hungary and the Netherlands there was a considerable share of new stakeholders attending the second workshop, which were mainly derived from the scenario specific second round stakeholder analysis, though it appeared hard to involve future stakeholders. It was also difficult to enrol public interest groups in both Hungarian workshops and governmental bodies in the second UK workshop. Furthermore, it was difficult to assure participation of all identified stakeholder categories that were distinguished within the five main groups. The low number of participants in the UK second workshop was due to a flu epidemic halving the number of participants. In the Netherlands the second workshop was organised together with a related project (Van Gaasbeek et. al., 2000). Furthermore, it can be noted that numbers were not the only way of evaluating the stakeholder involvement results. The share of different stakeholder groups, the quality of the involvement and process results were important too.

Table 4. Numbers of interviews and workshop participants per stakeholder group

	Interviews		*WS1 attendance*			*WS2 attendance*		
	Hun	NL	Hun	NL	UK	Hun	NL	UK
Total companies	*17*	*9*	*18*	*5*	*8*	*12*	*5*	*3*
• (Food) Service sector	1	2	2	1	-	2	-	-
• Appliances producers	3	1	-	3	1	-	2	-
• Retail/wholesale	2	2	2	-	3	1	1	2
• Processors/manufacturers	6	3	13	1	3	8	2	1
• Primary producers (agriculture, cotton, mining)	5	1	3	-	1	1	-	-
Total public interest NGOs	*1*	*9*	-	*3*	*5*	*2*	*2*	*3*
• Environmental groups		4		-	2	1	2	2
• Consumer groups		4		1	3	-	1	1
• Other	1	1		1	-	1	1	-
Total Government	*1*	*4*	*5*	*3*	*2*	*2*	*5*	*1*
• Ministries and National research councils		2	3	2	2		3	1
• Local/regional Government	1		1			1		
• Other (agencies)		2	1	1		1	2	
Research institutes and universities		*7*	*11*	*6*	*3*	*12*	*8*	*3*
Other (media, consultants)		*2*	*1*	-	-	-	*2*	-
TOTAL	**19**	**31**	**35**	**17**	**18**	**28**	**22**	**10**
Number of new stakeholders in the 2nd set of workshops						*15*	*6*	*1*

From a process point of view, it was also extremely important to keep stakeholders involved between the workshops, which was defined as stakeholder management (Quist, Pacchi and Van der Wel, 2000). Activities included the distribution of workshop reports, distribution of the constructed scenarios (the latter sometimes in combination with a questionnaire), further stakeholder interviews, participation in related meetings and projects partly as experts, while in UK even a SusHouse 2050 stakeholder club was initiated.

In general, involvement in the process and the two sets of workshops was quite well appreciated by participating stakeholders. This can for instance be illustrated by evaluation remarks made in the UK workshops that the workshops brought together organisations from the food supply chain that had never met or exchanged views before. However, evaluation of the second workshop showed that especially the (ambitious) implementation and stakeholder co-operation objectives were not fully met. Follow-up activities appeared to be necessary to support this. Section 3.3 will elaborate further upon this.

3.2 Sustainable Scenarios and Assessments

Based on the results of the first set of stakeholder creativity workshops that included numerous ideas on the sustainable function fulfilment and proto-scenarios elaborated by the participants, the research teams in the three countries developed scenarios depicting future sustainable ways of SCE, using the Design Orienting Scenarios (DOS) methodology developed by Manzini and Jegou (2000). DOS type scenarios contain a 'vision' describing the general atmosphere and the main characteristics. Additionally, this type of scenarios includes 'proposals' for product-service systems that contribute to the realisation of the sustainable function fulfilment, while they also contain a 'snapshot' description of living in that particular scenario in the year 2050 encompassing all the proposals in that specific DOS. Table 5 summarises the Design Orienting Scenarios (see also Green and Young, 2000; Quist, 2000; Toth et al., 2000; Young, 2000; Young et al., 2001). Three scenarios were developed in each country with two being similar for the three countries and one being unique to that country.

The scenarios were assessed for environmental improvement, economic changes and consumer acceptability. The *environmental assessment* used a system analysis approach with indicators, such as energy requirement, water usage, transport, domestic waste production and pesticide usage to assess if the scenarios would achieve a Factor 20 reduction in household environmental impacts. Briefly, the assessment revealed that most scenarios would reduce the environmental burden considerably. However, quite surprisingly, the 'Virtual Shopping' scenario did not reduce the energy requirement, while for the Super-Rant scenario in the Netherlands it was found that, using energy data from present restaurants, the energy requirement would even increase. It was concluded that better data would be necessary, and that there is a huge potential for environmental improvement in the Dutch food service sector.

Table 5. Scenarios developed from ideas developed in stakeholder workshops in 3 countries

Design Orienting Scenarios	Hun	NL	UK
Local and Green Food is supplied from local, organic sources. People eat in street corner eating-houses, or purchase food in local street corner shops to prepare and eat at home.	X	X	X
Hi-tech Eating The household contains various hi-tech pieces of equipment for storing, cooking, eating and wastage, e.g. 'smart' cooker that either reads the 'smart' packaging on the food or scans the bar code from the bulk food packaging for cooking information and an intelligent eco-efficient fridge. Water and energy is re-used as much as possible in the house. All appliances are either rented because they are so expensive or, if they are bought, the appliance's lifetime energy has to be bought at the same time.	X	X	X
Hi-tech Rural Gardens The majority of the rural households and the city households have hobby gardens suitable for food production. The household's daily routine involves working in the kitchen garden from spring to autumn. Household members can do this due to increased spare time, flexible working hours and tele-working. Food production in the garden is connected with recreation, high-tech garden equipment. In case of rural households the surplus goes to the market.	X		
Neighbourhood Food Centre, renamed as Super-Rant At diner meal times households members go to the Neighbourhood Food Centre or Super-Rant that combines elements from the 20th century supermarket and restaurants Here they can have an individual catering subscription to the neighbourhood cook, they can eat together or can take their meal home. In the Supermarket part of the Super-Rant people can do their food shopping. Many people have put away their kitchen, only fridge, microwave, water cooker and coffee machine are left.		X	
Virtual Shopping Food is selected and purchased using 'interactive' Net shopping in virtual reality from the household. The food is delivered to either direct to the household or to the household's local 'street distributor'.			X

The *economic assessment* used a questionnaire to assess each scenario for economic aspects as competition, employment and the opportunities, threats and effects for different sectors and stages of the supply and consumption chain. It was found that the Local and Green scenarios would require huge economic changes, while the other scenarios mostly required moderate changes. Finally, the *consumer acceptance analysis* used consumer focus groups to evaluate the acceptability of the scenarios to consumers and to identify adopter profiles, while they were also asked about their likes and dislikes, outstanding features and suggestions for improvement of the scenarios. Consumers appeared to be very capable of giving useful and interesting feedback on scenarios, while they were also very capable of rating scenarios and scenario elements on applicability in their present daily life and their personal preferences.

3.3 Implementation workshops and action agendas

The second series of stakeholder workshops focused on implementation and strategies, stakeholder feedback on results, development of concrete implementation proposals, the development of stakeholder partnerships around proposals, policy recommendations and integrating results into an overall 'action' or policy agenda. A structured framework combining elements from strategy approaches and back-casting techniques was applied. Again the workshop structure in the three countries was rather similar, but there were small differences due to adjusting the general structure to local circumstances. On average, the second workshop was more vibrant and included increased stakeholder interaction, enthusiasm, support for results, increased consciousness, and additionally dissemination of results.

The workshop succeeded in obtaining interesting and useful feedback on the constructed scenarios, while both implementation proposals were elaborated and new innovative ideas were proposed. Furthermore, the unprocessed results contained the 'raw material' for policy recommendations and also for the construction of an action agenda. An example derived from the workshop in the Netherlands is shown in Table 6.

However, concrete stakeholder co-operation, new partnerships and alliances around proposals were scarce in the three workshops. Nevertheless, there was considerable interest from most stakeholders, including concrete intentions to continue the process aiming at follow-up, co-operation around concrete proposals and activities, and at imbedding of results in processes taking place in other networks. Additional activities after the workshops appeared therefore to be necessary for realising follow-up towards implementation.

Post-workshop activities could include small meetings for the formulation of policy recommendations concerning sustainable food consumption, scanning the results on 'knowledge' questions and chain impact, working on follow-up proposals together with stakeholders, etc. This was indeed done in Hungary and the Netherlands, while in UK unfortunately capacity was lacking. Results are dealt with in the next section.

Table 6. Results from the Dutch second SCE workshop

The subgroup working on the scenario DOS **High-Tech Eating** (in the Netherlands known as Intelligent Cooking and Storing) presented the following results and action points:

- Decreasing meat consumption by development and introduction of Novel Protein Foods (protein foods from non-animal sources). Important actors are companies (product development) and government (stimulation with help of instruments)
- Decrease of required cooling volume due to just-in-time delivery by local distribution systems and innovation in packaging (making storing outside the fridge possible)
- Combining distribution (and retribution) for different functions (food, cloths, laundry, mail etc.) with help of intelligent front door.
- International co-operation and agreement on standardisation (or making compatible) of kitchen and household food management systems.
- Making better consumer choices possible with help of advanced information systems also supplying chain information (source, environmental impact, animal welfare, etc.)

The second subgroup working on the scenario **Super-rant** (also known as the neighbourhood food centre) presented the following results and action points:

- Making the concept more sustainable based on the use of sustainable supply chains and environmental improvement compared to the present practice in the catering sector (with help of technology and behavioural changes).
- Stimulating social change is important for the implementation of the Super-Rant concept
- Combine functions in this concept where possible resulting in multiple space use for instance by using company or school canteens that are used after 5 o'clock as super-rant for the neighbourhood.
- Pilots by co-operating restaurants and supermarkets (with help of other local actors like housing associations and local authorities in new suburbs and the countryside.

The third subgroup working on the DOS **Local and Green** presented the following results and action points:

- Sustainable regional production and consumption requires strong changes in technology, culture, consumption patterns and behaviour (do it yourself). I requires support and activities by the government, companies, consumers and civilians and research bodies
- Education will be extremely important for required social change and teaching skills and knowledge on food.
- New appropriate technology should be put in modules that can be customised in the region.
- Biotechnology could offer opportunities to increase the local variety and could compensate for decreased variety in foods due to limiting import.

3.4 Follow-up, implementation and stakeholder co-operation

While the SusHouse project and the applied approach has been very successful in involving stakeholders in a process exploring sustainable futures and elaborating proposals fitting in sustainable future, it is interesting to investigate if further stakeholder co-operation and follow-up activities continued or occurred after the project.

So, to what extent were new partnerships created? In the UK there was some interest in further co-operation. After receiving the UK report on SCE, many of the stakeholders reacted positively and appeared interested in further discussion, opportunities for co-operation and developing related projects. Hence, there are in a sense seeds but no harvest in terms of post project stakeholder co-operation and partnerships. However, it has already been mentioned that the wide range of stakeholders present and articulating their views and ideas was very much appreciated by many participants.

In Hungary there were concrete plans for follow-up research, but funding appeared to be a barrier. Post-workshop activities led to a proposal for an international research programme focusing on the conditions and opportunities for implementation of the Local Green Diet scenarios in rural areas in several Central and East European countries. This research programme is still under development with members from the Hungarian research team strongly involved. There are indications that other partnerships have grown after the last workshop, especially among alliances of producers and distributors.

In the Netherlands there were a number of initiatives, but it proved also to be hard to get these funded. However, concrete co-operation developed between stakeholders around specific ideas and proposals. It led for instance to a workshop focusing on domestic appliances for treating Novel Protein Foods at home. It was organised jointly by a research body and a company, while numerous other stakeholders (many of them not involved in the SusHouse project) were recruited. After the workshop the organising parties developed a concrete research proposal on optimising kitchen appliances and food supply chains from an environmental point of view. It was supported by a number of stakeholders from different societal groups. Furthermore, the Dutch research groups originally involved in the SusHouse project developed also new proposals including a programme proposal for a transition towards sustainability in eating-out and the food-service sector. It has been nominated for funding and aims at bringing together innovative parties from the food-service sector in the Netherlands, while it also requires participation by public interest groups and research bodies having complementary expertise (food supply chains, catering sector, demand side). Finally, results and conclusions from the SusHouse SCE function were used as input for the policy process of the Dutch Ministry of the Environment on sustainable food consumption.

4. CONCLUSIONS

This chapter has illustrated the stakeholder oriented approach developed by the SusHouse project which used scenarios and workshops as a new method for involving stakeholders in sustainability issues and generating new stakeholder partnerships and alliances. The results of the application of this approach in Hungary, The Netherlands and the UK for Shopping, Cooking and Eating illustrate that stakeholders were willing to participate in two rounds of workshops aiming at bringing about sustainable function fulfilment. They developed numerous ideas for sustainable shopping, cooking and eating in the far future, contributed to the construction of scenarios depicting sustainable function fulfilment, and eventually to the development of concrete proposals, research questions, policy recommendations and action agendas integrating different kind of activities.

An important effect of the SusHouse project was the interaction between stakeholders from both different stages of the supply chain and different societal groups. These stakeholders were brought together, often for the first time, to generate sustainability ideas and later on policy recommendations and implementation proposals. The effect on these stakeholders should not be underestimated because many had not thought about environmental issues and certainly not for the whole supply-chain or in relation with the consumption system. Many stakeholders felt they had learnt from the interaction with other stakeholders especially concerning the potential for changes to the supply chain for environmental improvement. In addition, constructive interaction between stakeholders from different societal groups having different mental frameworks and values also leads to an exchange of opinions and visions. Such an interaction may lead to what other authors call higher order learning (on the level of values and opinions) and may even influence values and mental frameworks of individuals and stakeholders. This type of learning processes are essential and an important condition for sustainable development, not only on a societal, but also on an organisational and an individual level. We have indications that the interactive construction of scenarios or future visions contributes significantly to stakeholder learning processes and stakeholder support.

Concerning partnerships for sustainability, we have argued that very often new partnerships and alliances are necessary and that this can include partnerships and alliances between stakeholders from different societal groups. Our results include examples of such partnerships in which often one stakeholder has taken the lead. However, our ambitions were higher and it is necessary to emphasise that establishing new alliances and partnerships, especially between stakeholders having different mental frameworks (for instance when they are from different societal groups) or when dealing with long-term developments requires incubation periods and involvement processes of considerable time. Bringing relevant stakeholders together in transparent participatory processes focusing on the broader context of sustainability issues and on specific ideas and contributions through the exploration of sustainable futures - for instance through scenarios - is a potentially strong instrument and can be seen as a condition for realising partnerships, especially if consciousness and knowledge with respect to environmental issues is rather low or if there is not a strong feeling of urgency. It enables individual stakeholders not only to explore sustainable future alternatives including their potential interest and opportunities but also to interact and exchange freely with other relevant stakeholders from other societal groups. This facilitates broader learning processes especially important when it comes to the stage of partnerships, co-operation and alliances where agreements must be found around proposals and activities related to short-term follow-up and opportunities.

A number of new partnerships did develop, but this part of the applied methodology could be improved. The question then is *how* to improve it. First of all, one needs to ensure that the right stakeholders are involved and that these stakeholders send the right delegates. Many studies point also out that a champion or leader is necessary in partnerships or alliances in order to keep the process going. There may be a tension between the long-term vision and short-term interests of stakeholders. For instance, companies are known to invest only if returns or spin-off results are short term. Some extra funding might be necessary to bridge this gap, as lack of funding proved to be a major bottleneck in our case. To overcome these problems governments could take (part of) the risks related to future oriented activities and research. Interestingly, the programme on Sustainable Technological Development (Weaver et al., 2000) using a strongly related approach had also to deal with the issue of follow-up and implementation activities, and it faced rather similar difficulties with respect to funding and champions (Coenen, 2000). Another point not yet solved is if and how the full action and research agenda resulting from this type of participatory processes should be facilitated when the initial project is finished.

Finally, could the methodology discussed in this chapter have a broader relevance? In our opinion, this question must be answered positively. Concerning the possible broader relevance and applicability, this chapter has presented a workshop organisation and stakeholder management oriented methodology which has the potential for applications in other situations to explore not only sustainable fulfilment of household functions, but innovations towards sustainability in general. It is our opinion that, especially when improving the implementation and follow-up aspects, this methodology exploring future scenarios through stakeholder workshops and scenario assessments has strong potential in areas like the development of sustainable business, issues of local sustainability (for instance as an extension to present Agenda 21 approaches) and industrial ecology.

REFERENCES

Coenen, L. (2000) *DTO Illustratieprocessen, wordt vervolgd....?* (STD Illustration processes, to be continued....?) Report DTO-KOV011 (in Dutch), Delft: STD-office.

Ehrlich, P.R., J.P. Holdren (1974) Impact of population growth, *Science* 171:1212-1217.

Green, K., Ph. Vergragt (2001) *Towards Sustainable Households: a methodology for developing sustainable technological and social innovations*, paper at the 9th International Conference of the Greening of Industry Network, Bangkok January 21-25.

Green, K., J. Quist, K. Szita Toth, L. Toth and W. Young (1999) *Scenarios for Shopping, Cooking and Eating in 2050: a three-country study*, paper at the 8th International Conference of the Greening of Industry Network Conference, Chapel Hill.

Green, K., W. Young (2000) *The Shopping, Cooking and Eating Function*, Final Report, SusHouse Project, Manchester: School of Management, UMIST, Delft: Delft University of Technology, TBM Faculty.

Manzini, E., F. Jégou (2000) *The Construction of Design Orienting Scenarios*, Final Report, SusHouse project, Politecnico di Milano University, Interdepartmental Research Centre, Innovation for the Environmental Sustainability – Department of Industrial Design (CIR.IS-DI-Tec), Delft: Delft University of Technology, TBM Faculty.

Quist, J, K. Szita Toth and K. Green (1998), Shopping, Cooking and Eating in the Sustainable Household, paper at the 7th International Conference of the Greening of Industry Network, November 15-18; Rome.

Quist, J., M. Knot, W. Young, K. Green, P. Vergragt (2001) Strategies Towards Sustainable Households Using Stakeholder Workshops and Scenarios. *Int. Journal Sustainable Development* 4, 1: 75-89.

Quist, J., C. Pacchi, M. Van der Wel (2000) *Workshop Organisation and Stakeholder Management*, Final Report, SusHouse Project, Delft University of Technology, Technology Assessment Group (Nl) / Avanzi, Milano (I), Delft: Delft University of Technology, TBM Faculty.

Quist, J. (2000) *Towards Sustainable Shopping, Cooking and Eating in the Netherlands*, Final Report, SusHouse Project, Delft: Delft University of Technology, TBM Faculty.

Roome, N. (2001) Editorial – Conceptualizing and Studying the Contribution of Networks in Environmental Management and Sustainable Development, *Business Strategy and the Environments' Special Issue Networks*, Environmental Management and Sustainable Development, 10, 2: 69-76.

Tansey, G., T. Worsley (1995) *The Food System*. London: Earthscan.

Tóth, K.S., L. Tóth, Z. Szekeres, L. Szűts, Z. Galbács, J. Fenyvessy (2000) *Shopping, Cooking and Eating, Hungary*, Final Report, SusHouse Project, Szeged University College of Food Industry, Department of Food Technology and Environmental Management, Delft: Delft University of Technology, TBM Faculty.

Van Gaasbeek, A., M.J. Meeusen-Van Onna, G. Wiersma, K.J. Kamminga, H.C. Moll (2000) *Consument, Voeding, Milieu* – final report, DTO-report no. KOV015, Delft: programme for Sustainable Technological Development.

Vergragt, P., M. Van der Wel (1998) Back-casting: An Example of Sustainable Washing, in *Sustainable Strategies for Industry*, N. Roome ed. 171-184, Washington DC: Island Press.

Vergragt, P. (1999) *The Sustainable Household: Technological and Cultural Changes*, Proceedings of the 1999 International Sustainable Development Research Conference: March 25-26, Leeds, UK. Shipley, ERP Environment 276-381.

Vergragt, P.J., L. Jansen (1993) Sustainable Technological Development: the making of a long-term oriented technology programme, *Project Appraisal 1993*; 8 (3):134-140.

Vergragt, Ph. (2000) *Strategies towards the Sustainable Household*, Final Report, SusHouse project, Delft University of Technology, Department of Industrial Design, Design for Sustainability Group, Delft: Delft University of Technology, TBM Faculty.

Weaver, P., L. Jansen, G. Van Grootveld, E. Van Spiegel, P. Vergragt (2000) *Sustainable Technology Development* Sheffield: Greenleaf Publishers.

Young, W., J. Quist, K. Toth, K. Anderson, K. Green (2001) Exploring Sustainable Futures through 'Design Orienting Scenarios': the case of Shopping, Cooking and Eating, *Journal of Sustainable Product Design* (accepted for publication).

Young, W. (2000) *Shopping, Cooking and Eating, the United Kingdom*, Final Report, SusHouse Project, Manchester School of Management, UMIST, Delft: Delft University of Technology, TBM Faculty.

Young, W., J. Quist and K. Green (2000) *Strategies for Sustainable Shopping, Cooking and Eating 2050 – Suitable for Europe?* Proceedings of the 2000 International Sustainable Development Research Conference: 2000 April 13-14, Leeds, UK. Shipley, ERP Environment, pp. 430-437.

Chapter 16

CONCLUSIONS

The Prospects of Collaboration

ARNOLD TUKKER

THEO DE BRUIJN

1. INTRODUCTION

We started this book by noting that sustainable development has become the central perspective in environmental strategies around the world. As a result, those strategies are now aiming at transformation rather than simple environmental reforms. Throughout this book, the authors addressed the role of partnership and leadership as an instrument to engage society in such a transformational process in a variety of cases and circumstances. Although collaboration has become the new paradigm in industrial transformation, there is still much uncertainty about the workings and outcomes of collaborative strategies. When reading the chapters in this volume, it becomes clear, that the theoretical concepts describing how innovation processes leading towards sustainability work, let alone which role partnership and leadership can or should play in these processes, are far from mature. In this chapter, we will therefore make an attempt to organise the information given in this book into a more structured, and more coherent framework, before assessing the experiences and lessons of the cases. We will touch upon the following three issues. The first issue concerns the classification of the innovation processes leading to sustainability. The second issue concerns the 'management' of the different types of innovation processes. Which are the most relevant factors that influence the direction of

T.J.N.M. de Bruijn and A. Tukker (eds.), Partnership and Leadership, 295–313.

these processes? The answer to this question can then be linked to the third issue, which is the core question of this book: how can partnership and leadership play a role in the different types of innovation processes? Which functions can it have? In our introductory chapter we identified collaboration as a premise for sustainable development. At the same time it may block more radical innovations. In this chapter we will take stock of the experiences so far after having provided a structure for classifying different types of partnership and leadership. On the basis of all experiences gathered in the cases in this volume, learning points can be identified on how partnership and leadership for sustainable innovations can best be organised.

The set up of this chapter is as follows. In section 2 we will give our attempt to classify innovation processes in relation to partnership and leadership. Section 3 will analyse which types of innovation processes actually have been covered in this book. Then, the following section will address the learning points per category of innovation. We end with a general conclusion in section 5.

2. PARTNERSHIP AND LEADERSHIP IN RELATION TO INNOVATION

2.1 Incremental and radical innovations

The first issue to address is: if we talk about innovation processes leading to sustainability, what do we mean? A notion that in our view is a first useful tool to structure the information in this book, is that of the complexity of the innovation. In many chapters in this book, in some cases only implicit, a difference is made between relatively 'simple' innovations like 'add-on technologies', innovations that can be realised and implemented by a relatively limited set of actors, and more complex innovations. Various authors made a difference between incremental improvements, roadmaps and radical innovations (e.g. Stevels, 1999), optimisation, re-design and system innovation (Weterings et al., 1997; RAND, 1997), 'normal' innovation versus Industrial Transformation (IHDP Industrial Transformation Science Plan, 1999), compliance, beyond - compliance and sustainable enterprises (Roome and Cahill, 2001), sustaining versus disruptive technologies (Christensen, 1997) or eco-efficiency and systems change (Schot et al., 1997). Here we choose to differentiate between system optimisation, singular innovations, and system innovations (Butter et al., 2000). The main difference is the extent to which the existing socio-technical web is changed upon implementation of the innovation.

Optimisation leaves the existing production/consumption structure largely as it is. A reduction of environmental pressure can be reached by a relatively small, usual technical adjustment of the system or an end-of-pipe measure. They usually can be implemented on a relatively short time horizon, but the degrees of freedom to reduce environmental pressures are limited. *Singular innovations* can be defined as changes in socio-technical arrangements that touch upon a limited number of actors in the production-consumption chain. Innovations are usually process-integrated or do require the 're-design' of a part of the production-consumption chain. This usually takes more time, but with the larger number of degrees of freedom, higher reductions of environmental pressure are possible. *System innovations* in principle focus on societal functions and their related systems. They have the largest degree of freedom to reduce environmental pressure, but also need the alignment of activities of a large number of actors in society during a considerable time period. The time horizon of such changes is often decades.

Figure 1 puts the three innovation levels in perspective. In fact it concerns a representation of the well-known S-curve from innovation theory, where the performance of a system can be considerably improved for a while by optimisation and re-design, then reaches an inherent performance threshold, that only can be surpassed once a new system breaks through. In our introductory chapter we cited Roome (2001) who stated that in moving up the innovation hierarchy (from context-taking to context-breaking in his terminology), the boundary of the actors involved increases rapidly. System innovations that consciously aim at sustainability therefore are likely to require complex, collaborative settings. The classification of cases in our book was based on the assumption that the broader the partnership, the better they would be equipped to deal with system innovation (i.e. one would expect in Part III more successful cases of system innovation than in Part II). In section 4 and 5 we will come back to this issue and show that the reality is more complex.

Several authors in this book have referred to work of Ehrlich and Holdren (1971) to underpin the need for radical innovations. With the doubling of the world population, and the need for particularly the developing countries to increase the wealth per person considerably, one faces an estimated potential wealth growth of a factor 10 in one generation. For environmental researchers it has become almost an axiom that this has to be compensated by at least a factor 10 reduction in environmental pressure per consumption unit, and that system innovations directed towards sustainability are the only way to reach such reductions.

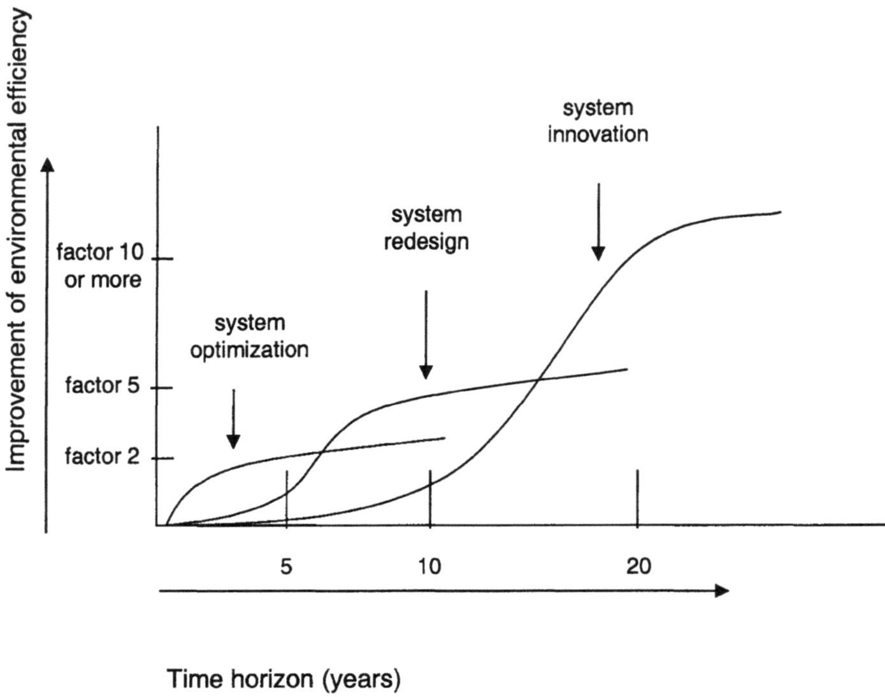

Figure 1. System optimisation, re-design and innovation (RAND, 1997; Weterings et al., 1997)

2.2 A multi-level perspective on innovations

How can innovation processes be modelled that bring system innovations nearer? The cases in this volume show that such changes are often hard to reach. It is the institutional context in which an innovation takes place that restricts the scope of innovation. Institutional adaptation can be very difficult since existing institutions represent long-standing habits of doing things. The possibilities for change are therefore path dependent. Change is only expected if and when the required change is not too big. Here the so-called logic of appropriateness comes in: When there is a large misfit, existing institutions will resist change (cf. March and Olsen, 1989). The required change then completely counters existing ideas, working routines, existing structures, competencies, etc. Only in cases where there are what is called medium adaptational pressures (so it is not a real stretch), change is possible, and this change will almost by definition then be incremental. Indeed, non-incremental innovations more often than not have their start *outside* the dominant institutional context (e.g. Christensen, 1997). A useful perspective on innovations therefore needs to address the context in which innovations take place. For this we will draw upon a model that has its roots in socio-technical change theory, and which was developed and articulated most eloquently by authors such as Schot, Geels and Kemp[1] (e.g. Schot, Geels and Kemp, 2000; Rotmans et al., 2000). These authors have elaborated on evolutionary economics that, simply stated, sees technological development as a process of variation and selection. The interaction between variation (determined by heuristics in research and development) and selection (by e.g. market forces and institutional and societal factors) leads to certain preferred trajectories (cf. Nelson and Winter's notion of technological regimes and Dosi's notion of technological paradigms (Nelson and Winter, 1982; Dosi, 1982). However, unlike Dosi and Nelson and Winter, who emphasised the knowledge-related aspects within a community of developers and engineers, the authors we refer to have broadened the notion to one of socio-technical regimes. By this, they acknowledge that it is not only designers and managers, but also users, policy makers, suppliers and societal groups who together 'determine' the boundaries for change and the direction of the regime. System innovations therefore require multiple stakeholder collaboration.

Next to the regime, which can be seen as a development at meso-level, *niches* can be discerned. These niches are relatively limited areas, in which new technologies can be developed and tested under relatively protected circumstances. Examples are specific market niches (where protection is provided by a specific market demand), or technological niches (where a specific actor network is willing to create the financial and other boundary conditions that make experimenting and learning possible).

The last (macro-)level has been called the *socio-technical landscape*. These are a set of fairly stable factors that are in principle external to the regimes and niches, and can only slowly be influenced. Examples include the existing material infrastructure, but also the existing culture, life-style, demography, or trends therein, etc. The landscape basically 'channels' the direction of the socio-technical trajectories (see Figures 2 and 3).

For the discussion in this chapter, we will consider system innovation as the equivalent of a change of a socio-technical regime. Niches play a key role in such changes, since they can act as stepping stones for maturing and the diffusion of radical new technology (cf. Christensen, 1997). Research on how these processes actually work and how a stimulating policy can be

[1] See also chapter 2 of this volume by Hartman et al..

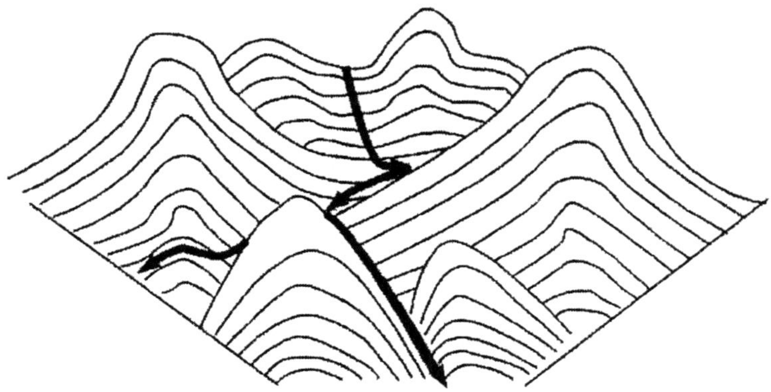

Figure 2. Topography of a socio-technological evaluation (Sahal, 1985:79, as reproduced in Geels and Kemp, 2000)

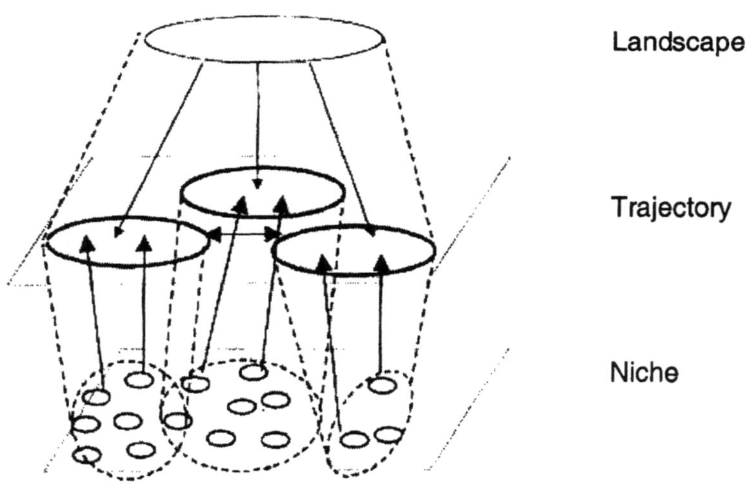

Figure 3. A multi-level perspective on innovation (Geels and Kemp, 2000:17)

developed is still in its initial stages, but the following tentative ideas have been formulated (Geels and Kemp, 2000):

- Successful transitions roughly can be divided in a preparatory phase (during which various singular innovations are tested in niches), a take-off phase (during which the alternative system gains some ground in its total form), an acceleration phase (where the new system starts to overtake the old system) and a saturation phase.

- A change in socio-technical regime usually tends to grow out of a combination of niche-technologies, and the form of change usually cannot be predicted.
- Often, a process of niche-cumulation can take place (i.e. that several key elements of a new socio-technical regime are applied in several, parallel or consecutive niches).
- Often, hybrid forms of old and new technology have a role as intermediates. This means that changes often originate outside the existing regime in niches, that subsequently grow to replace the old system. Therefore, it is important to link niches to existing technologies.

The whole idea behind this model is that system innovation, or the (normatively desired) change of a socio-technical regime, is a journey of which the direction roughly is known, but the exact end-point nor the exact roadmap not. The time horizon and the complexities are simply too large. The model also shows that system innovations in fact are the result of many small steps. Revolutionary changes starting from within the existing, dominant system are extremely seldom. What makes an innovation into a radical one is foremost a matter of changing the direction of change, rather than the magnitude of the change itself. Niches play an important role in this respect. In niches alternative technologies or solutions can be developed and tested that later on the trajectory may change the direction of changes in the dominant regime.

2.3 Actors in the socio-technical network

Given that a conscious pursuit of (system) innovation is a multilevel process, many different actors at various levels will be involved. System innovation is therefore not only multilevel, but also multi-actor. When we talk about partnership and leadership directed to sustainable change, who are the parties? As we have stated before, the answer to this question depends upon the intended level of change. Roome (2001:3; based on Roome and Cahill 2001) states in this respect that "moving up the innovation hierarchy increases in the complexity of the issues, the numbers of actors involved in change, and the number of linked, multiple technological and social options, the innovations and new practices that need to be undertaken and the uncertainties that have to be considered". Where companies can interact mainly with regulators, its usual customers and neighbours during the phase of compliance, sustainable enterprises need the involvement of all kinds of actors, probably from other than their usual production and consumption systems[2]. In a situation of singular innovations the range of actors might be limited to actors in the value chain. In system oriented changes multiple actor collaboration is required. In principle, the actors that are involved in maintaining a socio-technical network, all do so from a specific perspective. Each actor has certain interests, leading to positions and roles, of which the latter in turn leads to actions. In each socio-technical regime one sees involved as actors citizens/consumers, public authorities, industry, and a research and development infrastructure. They each have different interests, positions and roles (often also differing between members of one group), and hence a set of actions develops that may or may not be aligned when it comes down to pursue sustainable changes of the regime at stake.

For policy making, the findings in this and the previous section have various implications. We indicated that it is most probably impossible to define a clear end-goal or a road map for a factor 10 transition. We also indicated that in many cases it couldn't be expected that there is a natural tendency of societal actors to 'align' their actions in such a way, that they will co-operate smoothly to realise a transition. And finally we have to acknowledge that most Western societies have become network societies, making any notion of a powerful, central actor that can 'steer' other actors in a process of change obsolete.

In sum, a system innovation seems a journey without a roadmap, to a goal that cannot be specified up front, to be undertaken by a group of actors who all have their own ideas and positions, and of whom none has enough power to steer all the other actors involved. Any leadership should therefore take the following lessons into account. Particularly in the initial stages of a transition, stimulating socio-technical diversity and learning by doing is to be preferred above a deterministic, roadmapping approach. At the same time, a common idea about the rough direction(s) into which the transition could go should be clear. Therefore, such innovations need leadership next to collaboration. Stimulating experiments in new markets with a new selection regime (niches), and 'hooking on' to existing dynamics can be a useful approach as well. Only in the later stages of the transition a more selective and target-oriented approach could be tried (cf. Rotmans et al., 2000; Butter et al., 2000).

[2]As referred to earlier, Roome and Cahill (2001) distinguish between 'compliance', 'beyond-compliance' and 'sustainable enterprises'.

2.4 Implications of the framework

From the above, a number of conclusions can be drawn.

For *optimisation* and often also *re-design* the existing socio-technical trajectory in principle is not changed. The number of actors that needs to undertake action or change its mode of operation is limited. Most of the actors can continue in their existing role. The function of partnership and leadership at this level is clear, and also relatively simple. The 'leaders' in the network are those who are best capable of identifying the win-win situations that optimisation and re-design usually provide, and developing and implementing the related options usually implies a (limited) form of co-operation and partnership.

System innovation, however, demands a change of the socio-technical trajectory. Here, in the final situation it is likely that whole new actor constellations have been formed and/or that most actors have undergone a considerable change in their role and activity. Also, the actors involved in the partnerships needed, and the nature of their actions, can change dramatically during the process. Here, partnership and leadership is much more challenging. For setting out the overall *strategy* for change, leadership implies: being capable of building a common view among the actors/partners involved, envisioning the rough direction of the desired change of the socio-technical trajectory, and also being capable of (co-)organising the process that stands a chance of leading to such a change. Hartman et al. called this collaborative leadership (see chapter 2 of this volume).

For developing *niches*, probably relatively small actor networks that differ per niche will be sufficient. Usually, these actor-networks do not necessarily consist of actors that make up the network of the existing socio-technical paradigm (cf. Christensen, 1997). These niche-related networks are action-oriented, often lead by a company that was able to see the promising prospects of that niche, and are directed towards the development of the niche-solution (cf. Laws et al., 2001). However, it is also clear that the interaction between the strategy- and niche-level is a precondition for success: otherwise no niche-cumulation, diffusion, and transition will take place.

The conceptual framework we have chosen to use here combines many of the insights that our conceptual chapters offer. Hartman, Hofman and Stafford clearly point at the danger that – despite all benefits compared to a command-and-control approach – consensual approaches tend to favour incremental change, since they build upon existing socio-economic structures. Also, the emerging partnerships between NGOs and industry usually have to combine market and environmental objectives, which within traditional boundary conditions tends to favour incremental change. Co-operation and voluntary action, particularly towards more radical changes is often dependent on external threats (e.g. government regulations, crises, etc.). Poncelet's paper summarises this in one sentence: the search for the win-win. He postulates that less success can be expected from partnerships in situations where divergent interests, entrenched conflicts, and socio-economic inequality prevails. Furthermore, Georg and Irwin demonstrate the multilevel character of innovations by showing that the contrast between local and global action is not as clear as often thought, and that the most interesting phenomena are actually taking place at the interface between local (read: niches) and global (read: overall strategy and knowledge, and the relation with changes on the above-local level). Finally, our framework sees innovations, be it incremental or radical, as step-by-step developments. Taken together with the notion that system changes cannot be planned in a deterministic manner, the need for continuous learning along the process of change is crucial. Both the chapters by Roome and Clarke, and by Andersen underscore the need and importance of learning. De Bruijn and Hofman also stress the need for parties to engage themselves in learning processes. One-off improvements are in the end of limited value.

3. CLASSIFICATION OF CASES

In this section, we will classify the case-studies within the structure given above. In principle, this classification is different than the one we used to distinguish Part II (Collaboration along the value chain) and Part III (Multi-stakeholder partnerships) of this book. The latter one was based on the simple expectation that the more complex or radical the innovation, the more partners have to be involved. The two classifications are not necessarily contradicting (on the contrary), but as we will explain later, some interesting conclusions can be drawn from their comparison. Here we will classify the cases in the following way:

1. Partnership/leadership primarily aimed at incremental change or redesign;
2. Partnership/leadership primarily aimed at creating niches related to potential system innovations.
3. Partnership/leadership aimed at arranging an overall process of system innovation.

Table 1 gives the result of this exercise. A short explanation:

- In most cases, the reasons for classifying a case as incremental/redesign, niche or system innovation are self-evident.

- The two cases concerning SME support networks (Hines, and De Bruijn and Hofman) mainly discuss the support networks themselves. However, the papers implicitly suggest that the improvements that would be stimulated would be at the incremental/re-design level mainly.
- The cases of Roome and Clark concern changes at individual company level, and therefore by nature have limitations. For that reason they have been classified as 'incremental changes'. Yet, applying the lessons of Roome and Clark will probably be a fundamental challenge for the firms at stake. This is not to say that firms with these capabilities will not embark on system innovation – maybe on the contrary, it is likely that the firms complying with Roome and Clark's description will at least be sensitive for such changes. However, as stated by Poncelet and Hartman et al. in this volume, it might need a materialisation of pressure by *external* events before also these firms might embark on more radical change that might imply a total re-positioning of their business.
- The case of Kruijsen concerning photovoltaic cells might have been classified as system innovation as well. In fact, her analysis focuses on the question under which conditions the present (niche) situation can be overcome and a broad diffusion can be reached.
- The two cases of Georg and Irwin are not completely of the same nature. The development of a CFC-free freezer probably can best be classified as re-design, whereas the ecological village combined so many different novel (though individually existing) solutions, that the overall result probably can be best described as a niche.
- The case of Turcotte and Ali concentrates on a stakeholder roundtable that discusses a novel approach towards chemicals management. However, the implications of implementing such a novel approach in practice will cause major changes in the chemical industry (a sunset of some sectors, and expansion of others), and therefore it has been classified as a system innovation.

It is telling that the bulk of the cases in our book have to be classified as re-design or incremental innovation. It has to be reminded that for this book the editors already made a selection from the much broader set of cases presented at the 1998 Greening of Industry conference in Rome, and that this selection represented the most interesting and far-reaching cases. This reflects probably the fact that the thinking, let alone the activities on how to reach true factor 10-20 reduction in practice are still in their infancy and that most experience is available with re-design and incremental innovations.

Table 1. Classification of cases as incremental innovations, niches for system innovations, and system innovations

Incremental changes/Redesign	System innovations - Niches	System Innovations – Overall process
Roome and Clark: influence on strategic planning of a firm as a response to demands to more sustainable practices.	Georg and Irwin: Development of an ecological village in Sweden and the development of a CFC-free refrigerator	Turcotte and Ali: Description of a multi-stakeholder roundtable that aimed at a virtual elimination of persistent, bioaccumulative and persistent substances (PBTs), which would imply structural changes in the chemical industry
Andersen: Several cases of interfirm learning in relation to the paper and pulp industry (e.g. closing the water system and a simultaneous change in usage of types of chemicals and the development of high quality recycled paper)	Maier: Introduction and development of organic food on the Swiss market by a partnership of Co-op (retailer), Bio Suisse (labelling organisation), and Luxfood (producer of convenience foods)	Quist, Green, Toth and Young: Application of a methodology aimed at reaching a factor 10-20 reduction in environmental pressure for shopping, cooking and eating.
Canning and Hanmer-Loyd: Four different cases: a) a product return system for used telecommunication equipment; b) packaging redesign and return systems for cables delivered to a telecommunications operator; c) return and recycling system for used plastic retrieved from an IT company's products; d) labelling of plastics with the aim of improved recycling	Kruijsen: Introduction of photovoltaic cells in the Dutch market	
Hines: Description of a business support network (mainly supporting SMEs in realising incremental changes/redesign)		
Fadeeva and Halme: Eight different cases into sustainable tourism. Main goal of the projects: improving quality of life in the neighbourhood, or dealing with direct effects of tourism		
De Bruijn and Hofman: Support networks for SMEs		
Hastings: Stakeholder consultation by Shell commencing natural gas production in Peru		

This is an important statement, in view of the dichotomy many authors, including our own analysis in this concluding chapter, have suggested: that re-design and incremental change is relatively painless, since it can be reached in existing actor constellations, and that system innovation is often not, since this is a form of 'creative destruction' where one system including its actors is (gradually) replaced by an other. Given the multilevel and multi-actor characteristic of system innovations these changes are also hard to realise and 'manage'.

In the next sections, we will discuss the lessons about partnership and leadership on each level distinguished in this book. We will also discuss the issue if broader partnerships are better equipped to realise system innovations or not.

4. LESSONS LEARNED BY PARTNERSHIP TYPE

4.1 Incremental innovations and redesign

In the conceptual part of this book, Roome and Clarke made an extensive analysis of the characteristics of individual firms that enable them to develop their capability for environmental management and leadership. In short, these are:

1. Predisposition to mutual learning and change with other stakeholder groups;
2. Openness and responsiveness to multiple perspectives;
3. Access to managers with highly developed networks, networking skills and capabilities in facilitating change through these networks;
4. Use of networks of stakeholders to inform, confirm and validate strategy;
5. High process skills of managers to reconcile problems related to differences of perspective and language of the (extensive) network;
6. Facilitation of input from multi-stakeholder networks at all levels (strategic and operational).

When we look at the cases that we classified as incremental change and re-design, we see that indeed the most successful ones comply with these preconditions. Andersen's cases position the paper firm Danisco Paper as a relatively small entity with limited research capacity, but who compensates this by maintaining extended networks with suppliers, competitors, and customers, and therefore ending up as very innovative. For Stora Dalum a similar analysis applies. Canning and Hanmer-Lloyd show similar success factors in their cases, such as trustworthy behaviour, backing of senior management, and a focus on information exchange and negotiation. Faveeda and Halme describe in their cases on sustainable tourism networks factors such as a long-term orientation, looking for steady small wins, and an organisation that is a combination of flexibility and formality and that is open for feedback for adjustment of the network strategy.

Yet, these cases also indicate the limits to an intra-firm leadership approach only. It is not likely that Stora Dalum had embarked on the risky trajectory of developing a closed-loop water system for their mills had there not been external pressure. Canning and Hanmer-Lloyd present various cases where the absence of threat to traditional sales and interests were described as a success factor. On a larger scale, Hines' description of the Wales Environmental Services Partnership shows that straightforward conflicting interests about access to funding and formal leadership can bring partnerships down. Hastings' case on the behaviour of Shell in developing countries also shows two sides of the same coin: at present, where local groups and public opinion can exercise power a responsible company, complying with Roome and Clarke's rules, but in former times just playing out their power as a monopolist. And De Bruijn and Hofman find that a strictly voluntary approach doesn't succeed in engaging laggards in pollution prevention. These findings confirm the statements by Hartman et al. in chapter 2 who stated that adversarial (and coercive) approaches may have to be employed simultaneously.

4.2 Niches for system innovation

Kruijsen (diffusion of PV cells), Maier (introduction of high-quality, organic food in a Swiss catering system) and to some extent Georg and Irwin (CFC-free refrigerator and ecological village) have described developments that could be the start of system innovations, but are currently mainly still in the niche phase.

It is quite clear that a certain amount of collaboration and particularly leadership are preconditions for niches to be successful. The cases indicate, however, that how the niches are created can differ quite a lot from case to case. Georg and Irwin show for the case of the CFC-free refrigerator that it took a number of very coincidental events, before an fruitful idea was on the table that really could be taken up by a leader – in this case there had been no active search for a more sustainable cooling apparatus.

Several blocking factors for breaking out of the niche stage can be found in these cases. It is striking, though, that there seems to be no single pattern or cause that determines success or failure. In the 'Greenfreeze' case, a classical problem almost killed this initiative. The innovative product was put on the market by a relatively small niche player, caught the main players by surprise, with as a result that the main players tried to use their power to close the market for this new product. Here, the powerful backing of Greenpeace and the related support generated effectively blocked this attempt. The large competitors quickly changed their strategy and started to develop 'Greenfreeze'-like freezers as well – effectively leading to a quick breakthrough of this technology to the market. In Maier's Luxfood case, a collaboration of just three market parties (food producer Luxfood, labelling organisation Bio Suisse and the powerful retailer Co-op) was enough to facilitate the introduction of biological convenience foods. In short, Co-op created the direct demand (expecting a need felt by final consumers), Bio Suisse provided the instrument of credibility, and this was for Luxfood a sufficient incentive to get into motion. However, mainly since the potential of a broader market introduction of biological convenience foods fell outside Luxfood's 'mental map', for Luxfood it remained a one-off affair. Here, the internal culture and strategy of one firm caused a less extended diffusion of biological convenience foods than what could have been the case. In Kruijsen's PV case a more fundamental blocking factor is at stake that cannot be solved by a single partner in the network. In this case, it is clear that a close co-operation between developers and users is needed for an optimal chance on success. However, the financial threshold still prevents large-scale diffusion of this technology.

Overall, the cases show that it is of the utmost importance to engage sufficient partners along the value chain. However, the cases make also clear that in order to create niches as such not necessarily a broad, multistakeholder process is needed. In various cases, a collaboration of just a handful of partners under targeted (and generally industrial) leadership was enough to get the niche off the ground (cf. Laws et al., 2001). However, the breakthrough out of the niche seems in at least some (though not all) of the cases a different matter. Where in Maier's case the culture of one actor that was part of the niche players formed the blocking factor for further progress, in the other ones the blocking factors could not be solved by the initiators of the niche only.

4.3 System innovation

The number of attempts to realise true system innovations described in this book is quite limited. One could see the multi-stakeholder collaborative roundtable described by Turcotte and Ali as one example. However, this attempt to realise a system innovation in chemicals management failed due to the fact that the long-term, basic goals of different key stakeholders could not be reconciled. The case of sustainable shopping, cooking and eating of Quist et al. provides the only systematic approach to reach system innovations presented in this book. Here, a group of researchers actually acted as intermediates and leaders by organising a process that included all other stakeholders. The approach was successful in creating initial co-operation, but the extent to which follow-up activities after the project were created with enough momentum is not clear. Given the long time horizon that system innovations requires this is a serious uncertainty.

It is clear that in these both cases collaboration along the value chain would have resulted in an even lower chance for a successful system innovation, as is the case in these multi-stakeholder situations.

4.4 Overall conclusion

So what does this imply for partnership and leadership? It is clear that, particularly when we look to industry, a number of characteristics can be given for companies that will make them relatively better (environmental) leaders than others. Such preconditions have been described by Roome and Clarke, and in fact are nothing more and nothing less than general preconditions that have to be met by firms if they want to be innovative, flexible, and learning organisations in general. If the world would exist largely of such ideal firms, probably the following goals could be reached easier than today:

- Broader implementation of environmental improvements at the level of incremental change and redesign;
- Sensitivity and an open eye for developments in the firm's environment that could lead to the need for system innovation.

The simple rationale behind this division is that in general consequences of incremental change and re-design are relatively simple to oversee, need a partnership and leadership relation of just a limited number of actors, and in general tend to *re-inforce* the competitiveness the network as a whole. System changes, however, often are a form of Schumpeterian 'creative destruction'. Here, one cannot expect in all cases that a network of relatively open-minded and eager-to-learn actors will start to move in the right direction without some form of external pressure. The process that has to be embarked upon is uncertain, and will – we repeat will – create losers, just like the big ocean liner companies were the losers once the system change in passenger transport to air traffic really took off. And, as shown in our book by both Hartman et al. and Poncelet, a network will only move without external pressure if a win-win is at the horizon – within a time frame reasonably acceptable in business terms (cf. Walley and Whitehead, 1994).

So where do we have to look for leadership for radical change? And does such radical change by nature need an (organised) multi-stakeholder collaboration to be successful? To start with the last question, the answer is no. As we saw in the last section, niches can be created by just a few imaginative and smart actors. Depending on the socio-technical setting, these niches can grow in power more or less autonomously, until they overcome and replace an existing system more or less 'naturally' – again, the ocean liner story given above is a good example (cf. Christensen, 1997). Of course in the growth stage of a niche more and more actors will become involved, but not in a planned way: it just 'happens'.

The challenge is of course that when it comes down to sustainable development, we cannot be satisfied with system innovations that 'just happen'. In that case the 'lottery of innovation processes' might provide us with system innovations which are not desirable from a sustainability viewpoint, and might keep innovations which should break through concealed in niches. Here, a multi-stakeholder process – if only in an analytical phase – seems inevitable, and leadership is essential. And here, after reviewing the cases in this book, it seems to us obvious that public authorities cannot stay passive. This does not mean that they should take up the leader role in all cases, but they cannot hide behind concepts like 'leaving it to the market'. The present markets, if left on their own, cannot be expected to steer innovation into a sustainable direction. This forms a strong legitimisation for a role of public authorities. It is obvious though that this is not an easy panacea. First, 'the public authorities' are far from monolithic. Second, particularly the more operational parts of the public authorities are linked to the existing socio-technical systems and might actually be part of the problem that opposes change. However, it is also clear that public authorities, ultimately, have a certain power to set the rules by which the market has to play (WBCSD, 2001). It might be that what has to be changed is made clear under another leadership, e.g. pro-active intermediate groups, companies or NGOs, but public authorities should at minimum be prepared to use their power in favour of change.

As we indicated earlier in this chapter, any leadership should take the following lessons into account. Particularly in the initial stages of a transition, stimulating socio-technical diversity and learning by doing is to be preferred over a deterministic, roadmapping approach. At the same time, a common idea about the rough direction(s) into which the transition could go should become clear. Stimulating experiments in new markets with a new selection regime, and 'hooking on' to existing dynamics can be a useful approach as well. Only in the later stages of the transition a more selective and target-oriented approach could be tried (cf. Rotmans et al., 2000; Butter et al., 2000). Of course, this journey has a much higher probability of success if the participants are open for learning, and form a partnership. However, we have to acknowledge that the thinking how to 'guide' transitions is still in its infancy[3]. To complain that many (environmental) leaders in industry just get stuck in incremental innovations and re-design is to some extent an injustice: within their 'natural' boundaries they do the maximum that can be expected. The real challenge is that humanity, maybe for the first time in history, now stands for the task to 'direct' the socio-technological trajectory it will follow in the future from a normative point of view. This has to be done in a time frame, where absolute power has become non-existent and the real power is in the hands of groups that together make up the network. Hence, after all Partnership and Leadership may have been the wrong title for this book. The real question for modern society now is how we can find a way to create a sublimation of these two elements: a Partnership, that *is* the Leadership.

REFERENCES

Butter, M., Th. Van Hoorn and A. Tukker. *Van duizend bloemen naar één boeket. Technologische innovaties voor milieu en economie.* (From thousand flowers to one bouquet. Technological innovations for environment and economy) TNO-STB report 00-37, Delft, the Netherlands.

Christensen, C. M. (1997). *The Innovator's Dilemma. When New Technologies Cause Great Firms to Fail.* Boston: Harvard Business School Press.

Dosi, G. (1982). Technological paradigms and technological trajectories. A suggested interpretation of the determinants and directions of technical change. *Research policy* 11: 147-162

Ehrlich, P. and J. Holdren (1971). Impact of Population Growth. *Science* 171: 1212-1217

Geels, F. and R. Kemp (2000). *Transities vanuit sociotechnisch perspectief.* (Transitions in socio-technical perspective). Research paper, CSTM/MERIT, Enschede/Maastricht, the Netherlands

International Human Dimensions Programme on Global Environmental Change (1999) *Industrial Transformation Science Plan*, IHDP Report No. 12, Bonn.

Laws, D., L. Susskind, J. Abrams, J. Anderson, G. Chapman, E. Rubinstein and J. Vadgama (2001). *Public Entrepreneurship Networks. The Environmental Technology and Public Policy Program*, Department of Urban Studies and Planning, MIT, Cambridge, US

Lindblom, Charles E. (1959) The Science of Muddling Through. *Public Administration Review* 19, 2: 79-88

March, James G. and Johan P. Olsen (1989) *Rediscovering institutions; The organizational basis of politics*, The Free Press.

Nelson, R. and S. Winter (1982). *An Evolutionary Theory of Economic Change.* Harvard University Press, Cambridge, US

[3]Indeed, it is telling that many authors flatly reject the idea that such major innovations can be 'steered' (cf. Lindblom, 1959)

RAND (1997). *Technologieradar*. RAND Europe for the Dutch Ministry of Economic Affairs, the Hague, the Netherlands.

Roome, N. and E. Cahill (2001) *Sustainable production - challenges and objectives for EU research policy, report of the expert group on competitive and sustainable production and related service industries in Europe in the period to 2020*, EUR 19880 Research Directorate General, Final report.

Roome, N. (2001) *Policies and Conditions for Environmental Innovation and Management in Industry*, paper presented at the International Conference Towards Environmental Innovation Systems, Garmisch-Partenkirchen, September 27-29.

Rotmans, J., ed. (2000) *Transities en transitiemanagement*. ICIS and MERIT, Maastricht, the Netherlands.

Sahal, D. (1985). Technological guideposts and innovation avenues. *Research Policy*, Vol. 14: 61-82.

Schot, J. (1997) Duurzame ontwikkeling, een uitdaging voor de industrie (Sustainable Development, A Challenge for Industry, in *Ketenbeheer* (Chain Management), K. Lulofs and G. Schrama, Enschede: Twente University Press.

Schot, J., E. Brand and K. Fischer (1997) The Greening of Industry for a Sustainable Future: Building an International Research Agenda, *Business Strategy and the Environment*, vol. 6, no. 3: 153-162.

Stevels, A. (1998). Eco-efficiency and sustainability at Philips Sound & Vision. in *Product Innovation and Eco-efficiency*, J.E.M. Klostermann and A. Tukker eds., Kluwer Academic Publishers, Dordrecht, the Netherlands.

Weterings, R., E. Smeets and G. Annokkee (1997) *81 Options for Sustainable Development*. Ministry of Housing, Physical Planning and Environment, Den Haag, the Netherlands

Walley, N., and B Whitehead (1994). It's not easy being green. *Harvard Business Review*, Vol. 72, No. 3.

World Business Council for Sustainable Development (WBCSD, 2001). *Sustainability through the Market. Seven keys to success*. WBCSD, Geneva, Switzerland

BIOGRAPHIES

Basmah Ali was an M.Sc. Student in Commerce at Concordia University. She graduated in 2001. She is now working in the placement department of a university in Zayed University, in Dubai, United Arab Emirates.

Maj Munch Andersen Ph.D. is currently employed as sustainability advisor at the Danish Ministry of Trade and Industry, working on developing a Danish green industrial development strategy. From 1993 to 1999 she has worked as a researcher at the Copenhagen Business School. Her research covers issues of corporate environmental behaviour, innovation and business strategy from an industrial dynamics perspective, with a special focus on interfirm relations, innovation systems and technological trajectories.

Louise Canning is senior lecturer in marketing at Bristol Business School, UWE Bristol. She held international sales and marketing posts in the steel and engineering industries for a number of years before joining UWE Bristol, where she obtained her Ph.D. Her research interests lie in the areas of business-to-business marketing and environmental management and has published in the *Journal of Business and Industrial Marketing* and *Business Strategy and the Environment.*

Sarah Clarke qualified as an electrical engineer and then took an MBA from Manchester Business School. She graduated with a Ph.D. from the Schulich School of Business, York University Canada. She has authored a number of papers on technology management for sustainable development. Sarah has been active in bringing ideas about sustainable development to the engineering profession and is involved in supporting the access to opportunities of minority and disadvantaged groups.

Theo J.N.M. de Bruijn Ph.D. is a senior research associate at the Center for Clean Technology and Environmental Policy (CSTM) of the University of Twente in the Netherlands. His research interests focus on the governance of industrial transformation. He teaches environmental management within several postgraduate courses and for under-graduates. De Bruijn is the European co-ordinator of the Greening of Industry Network. He was also a member of the Scientific Planning Committee of the International Human Dimensions of Global Environmental Change Program (IHDP) on Industrial Transformation. Among other editorial memberships he is a member of the

Editorial Board of the book series on Eco-efficiency & Industry with Kluwer Academic Publishers.

Zinaida Fadeeva MSc is a research associate and a doctoral candidate at the International Institute for Industrial Environmental Economics at Lund University, Sweden. Her research interests are in inter-organisational environmental management and sustainable tourism. She co-ordinates and teaches environmental management and cleaner production to masters' students and educates Eastern European industry on environmental management.

Susse Georg is Associate Professor at the Department of Organization and Industrial Sociology, Copenhagen Business School. She has published work on environmental regulation, the development of cleaner technology, the 'greening' of industrial policy. Her current research interests include the institutionalisation of environmental concern in organisations, environmental communication and the expressiveness of organisations, environment and governance, and environmental sociology.

Ken Green is Professor of Environmental Innovation Management at Manchester School of Management, UMIST, UK and Director of the Centre for Research on Organisations, Management and Technical Change (CROMTEC). His current research interests are in the business and policy issues raised by the 'greening' of production and consumption activity and the connections between technological innovation and sustainability. He works with the Greening of Industry Network, International Human Dimensions of Global Environmental Change's Industrial Transformation Project and the UK Tyndall Centre for Climate Change Research.

Minna Halme Ph.D. is an assistant professor at the Department of Management and Organisations at the Helsinki School of Economics. Her main research interests include developing environmental corporate cultures, actor network and sustainable development, environmental communication and marketing, sustainable tourism and sustainable forest industry. She has published in *Business Strategy and the Environmental*, *Business Ethics Quarterly*, *The Sustainable Journal of Management* and in other journals and edited collections. She teaches corporate environmental management to masters' students and business executives.

Stuart Hanmer-Lloyd is currently reader in marketing and head of the Centre for Research Services (CERES) at Gloucestershire Business School. He holds a B.Sc. (Hons) from City University, an M.Com from Birmingham

University and a Ph.D. from Cranfield University. His research interests lie mainly in the areas of business to business marketing and environmental issues in marketing. He has published in these areas in a number of journals including the *European Journal of Marketing, Journal of Marketing Management* and *Business Strategy and the Environment.*

Cathy L. Hartman, Ph.D. is an associate professor of buyer behavior at Utah State University. Cathy studies 'green alliances', collaborations between environmental stakeholders to pursue mutually beneficial ecological goals. Her expertise centers on the interpersonal dynamics and ideological dilemmas involved in the formation, implementation, and continuance of these partnerships. Cathy is active in the Greening of Industry Network (GIN), having served as a co-editor for a special issue of *Business Strategy and the Environment* (Vol. 8, No. 5) on the 7th International Conference and as a member of GIN's planning board. Her work has been published in marketing and business journals as well as the popular press.

Marilu Hastings is Director of the Policy and Economics Program at the Houston Advanced Research Center in The Woodlands, Texas. The Program focuses on organizational policy and practice for sustainable development, innovative economic tools for resource conservation, the role of social capital in sustainability, and integrating rigorous scientific thinking into discourse about sustainability. Marilu earned a Master of Public Affairs from the LBJ School of Public Affairs at the University of Texas at Austin and a Master of Business Administration from the Graduate School of Business at the University of Texas at Austin. She obtained a Bachelor of Arts in economics and political science from Duke University.

Frances Hines is the Research Manager of the ESRC Centre for Business Relationships, Accountability, Sustainability and Society (BRASS). The Centre based at Cardiff University is a collaboration between the Department of City and Regional Planning, Cardiff Business School and Cardiff Law School. Frances has been a researcher into business, environment and sustainability issues at Cardiff University for eight years, and lectures on these issues. She is completing her Ph.D. on changing styles of environmental regulation in East Central Europe, and gained an M.Sc. in European Environmental Policy and Regulation from Lancaster University in 1994.

Peter S. Hofman is a research associate at the Center for Clean Technology and Environmental Policy at the University of Twente. His research centers on the role of policy and innovation in transition paths towards more sustainable production and consumption. Peter is active in the Greening of Industry Network (GIN), having served as a co-editor for a special issue of *Business Strategy and the Environment* (Vol. 8, No. 5) on the 7th International Conference, as a member of GIN's planning board, and as a co-editor of Ahead of the Curve (with Ken Green and Peter Groenewegen, Kluwer Academic Publishers, 2001). He currently also holds a post-doctoral position, funded by the Dutch Scientific Council, at the Centre for Studies of Science, Technology and Society at the University of Twente in which he is involved in analysis of socio-technical change related to the energy system and the development of socio-technical scenarios.

Alan Irwin is Professor of Sociology at Brunel University, West London, UK. He is author of *Citizen Science* (Routedge, 1995) and *Sociology and the Environment* (Polity, 2001). He was co-editor (with Brian Wynne) of *Misunderstanding Science?* (Cambridge, 1996). His current research interests include the public understanding of risk, technology policy, science and governance, and environmental sociology.

Joanneke Kruijsen holds a M.Sc. in Industrial Design Engineering from Delft University of Technology. In 1999, she received her Ph.D. at Delft University of Technology from the Faculty of Systems Engineering and Policy Analysis. In 1999, Joanneke joined The International Institute for the Urban Environment as project manager. She co-ordinates large European projects that deal with innovation towards sustainability with a focus on stakeholder participation, communication of research, technology and innovation. For the European Commission, Directorate General Research she is acting as evaluator of research proposals. Joanneke is a member of the governing board of 'O2 Nederland', the Association for the encouragement of Sustainable Product Design.

Simone Maier studied business administration and economics at Witten/Herdecke University, Germany, and University of Haifa, Israel. She took part in a transdisciplinary research project involving nine research teams from various disciplines at different Swiss universities around the topic of sustainable development in the food sector. Her dissertation written in 2000/2001 at University of St. Gallen, Switzerland, deals with the strategic challenges arising from the introduction of organic products to the Swiss catering industry. Since summer 2001 she works as a strategy developer for Swiss Union of Raiffeisen Banks.

Eric Poncelet earned his Ph.D. in Cultural Anthropology from the University of North Carolina at Chapel Hill (UNC-CH), where his research concentrated on multistakeholder environmental partnerships in Europe and the United States. As a Visiting Scholar at UNC-CH, his publications explored the influences of cultural differences on the practices and products of multistakeholder collaborative processes. Since 2000, Dr. Poncelet has been working as a practitioner in the domain of collaborative environmental planning and management, where his emphasis has been on designing, convening, and facilitating collaborative decision-making processes. He has worked for the Natural Resources Leadership Institute and is now with CONCUR, Inc.—an environmental mediation firm specializing in consensus building and policy analysis.

Jaco Quist is a Lecturer in Technology Assessment at Delft University of Technology, Faculty of Technology, Policy and Management. His research focus concerns managing system innovations towards sustainability in a multi-actor context and as well as social aspects of technology development. Between 1998 and 2000 he worked on the EU funded project 'Strategies towards the Sustainable Household – SusHouse'. Before 1998 he was a project coordinator on the Dutch governmental research programme for Sustainable Technological Development.

Nigel Roome holds the chair in Sustainable Enterprise and Transformation at Erasmus University Rotterdam. He is head of the Erasmus Center for Sustainable Development and Management. He previously held chairs in the Netherlands and Canada, and academic positions in Britain. He is widely published on topics related to business and environmental management, sustainable development and global change, including - *Sustainability Strategies for Industry: The Future of Corporate Practice* - published by Island Press in 1998. He recently chaired a European Commission Expert Group on RTD&I policies for sustainable and competitive European production systems.

Edwin R. Stafford, Ph.D. is an associate professor of marketing strategy at Utah State University whose research centers on environmental collaboration and the diffusion of greener and cleaner technologies. Ed has worked with a variety of environmental NGOs, including Greenpeace, Ducks Unlimited, and the Nature Conservancy, examining their collaborative and market-based campaigns. Ed is active in the Greening of Industry Network (GIN), having served as a co-editor for a special issue of *Business Strategy and the Environment* (Vol. 8, No. 5) on the 7th International Conference and as a member of GIN's planning board. He also is a frequent op-ed contributor to the news media on corporate environmental issues.

Klára Szita Tóth is both educated as a chemist and an environmental engineer and has a Ph.D. in economics. Currently, she is an associate professor at the University of Miskolc, Faculty of Economics. Her research covers areas on sustainable households, scenarios for sustainable food chains and environmental assessments and Life Cycle Assessments of foods. She has published several books, chapters and many other publications in both Hungarian and English. Dr. Tóth is also a member of the Future Research Committee of the Hungarian Academy of Sciences, the Director of the Regional Cleaner Production Centre Szeged and is involved in several other committees concerning Food Sciences and Environmental Management of Food production.

Arnold Tukker Ph.D. graduated as chemist at Utrecht University in 1987. After three years with the Dutch Environment Ministry he joined TNO, where he is now a senior scientist/advisor at the Institute of Strategy, Technology and Policy (TNO-STB), TNO's strategic think tank in the field of innovation policy. He co-ordinates one of the about eight focal areas of TNO-STB, i.e. innovation policy for sustainability. In 1998 he published a book on societal disputes on toxic substances, which awarded him a Ph.D. from Tilburg University. He published some 20 peer-reviewed papers in various scientific journals. He edits the book series on Eco-efficiency & Industry with Kluwer Academic Publishers, and has positions in the editorial boards of the International Journal of Life cycle Assessment and the Journal for Sustainable Product Design.

Marie-France Turcotte is professor of strategy at the Université du Québec à Montréal. Her research interests pertains to corporate social and environmental responsibility, as well as the role of the civil society in corporate governance. She is currently working on research projects on the topics of multistakeholder initiatives and the socially responsible investing movement.

William Young Ph.D. is a Lecturer in Environment and Business at the School of the Environment, University of Leeds in the UK. He has research experience in environmental and sustainable development indicators for business, environmental reporting, sustainable production and consumption, sustainable households/lifestyles as well as ethical and fair trade. William Young is an UK Principle Expert sitting on British Standards Institute (BSI) and International Organization for Standardization (ISO). He is an Advisory Board Member for the *Eco-Management and Auditing Journal.*

Eco-Efficiency in Industry and Science

1. J.E.M. Klostermann and A. Tukker (eds.): *Product Innovation and Eco-efficiency.* Twenty-three Industry Efforts to Reach the Factor 4. 1997 ISBN 0-7923-4761-7
2. K. van Dijken, Y. Prince, T. Wolters, M. Frey, G. Mussati, P. Kalff, O. Hansen, S. Kerndrup, B. Søndergård, E. Lopes Rodrigues and S. Meredith (eds.): *Adoption of Environmental Innovations.* The Dynamics of Innovation as Interplay Between Business Competence, Environmental Orientation and Network Involvement. 1999 ISBN 0-7923-5561-X
3. M. Bartolomeo, M. Bennett, J.J. Bouma, P. Heydkamp, P. James, F. de Walle and T. Wolters: *Eco-Management Accounting.* 1999 ISBN 0-7923-5562-8
4. P.P.A.A.H. Kandelaars: *Economic Models of Material-Product Chains for Environmental Policy Analysis.* 1999 ISBN 0-7923-5794-9
5. J. de Beer: *Potential for Industrial Energy-Efficiency Improvement in the Long Term.* 2000 ISBN 0-7923-6282-9
6. K. Green, P. Groenewegen and P.S. Hofman (eds.): *Ahead of the Curve.* Cases of Innovation in Environmental Management. 2001 ISBN 0-7923-6804-5
7. J.B. Guinée (ed.): *Handbook on Life Cycle Assessment.* Operational Guide to the ISO Standards. 2002 ISBN 1-4020-0228-9
8. T.J.N.M. de Bruijn and A. Tukker (eds.): *Partnership and Leadership.* Building Alliances for a Sustainable Future. 2002 ISBN 1-4020-0431-1

KLUWER ACADEMIC PUBLISHERS – DORDRECHT / BOSTON / LONDON

MIX
Papier aus verantwortungsvollen Quellen
Paper from responsible sources
FSC® C105338

If you have any concerns about our products,
you can contact us on
ProductSafety@springernature.com

In case Publisher is established outside the EU,
the EU authorized representative is:
Springer Nature Customer Service Center GmbH
Europaplatz 3, 69115 Heidelberg, Germany

Printed by Libri Plureos GmbH
in Hamburg, Germany